Geometry Problems
One Step Beyond

Reuben Schadler

DALE SEYMOUR PUBLICATIONS

Illustrations: Bill Eral
Cover art: Julie Peterson

Copyright © 1984 by Dale Seymour Publications. All rights reserved. Printed in the United States of America. Published simultaneously in Canada.

Limited reproduction permission. The publisher grants permission to reproduce up to 100 copies of any part of this book for noncommercial classroom or individual use. Any further duplication is prohibited.

ISBN 0-86651-226-8
Order number DS01430

DALE SEYMOUR PUBLICATIONS
P.O. BOX 10888
PALO ALTO, CA 94303

7 8 9 10 11 12 13 14-MA-95 94

CONTENTS

PROBLEM SET 1
Warm-up	Name That Line	1
Problem 1	Stair Steps	3
Extension	Regional Parts	5

PROBLEM SET 2
Warm-up	Type Cast	7
Problem 2	Pentagon Search	9
Extension	Two's Company	11

PROBLEM SET 3
Warm-up	Alphabet Soup	13
Problem 3	Rain or Shine	15
Extension	Problem Child	17

PROBLEM SET 4
Warm-up	Fruit Stand	19
Problem 4	Afternoon at the Prevu	21
Extension	In the Balance	23

PROBLEM SET 5
Warm-up	Cubed Salad	25
Problem 5	Line Up	27
Extension	Surface Paint	29

PROBLEM SET 6
Warm-up	At the Vertex	31
Problem 6	Star Sum	33
Extension	How Many Sides?	35

PROBLEM SET 7
Warm-up	Half a Hex	37
Problem 7	As the Crow Flies	39
Extension	Size 'em Up	41

PROBLEM SET 8
Warm-up	Lighthouse	43
Problem 8	At the Ritz	45
Extension	Steamer Stripe	47

PROBLEM SET 9
Warm-up	Ricochet	49
Problem 9	Shortcut	51
Extension	Only with Mirrors	53

PROBLEM SET 10
Warm-up	Tangrams	55
Problem 10	Popsicle Stick Constructions	57
Extension	Right On	59

PROBLEM SET 11
Warm-up	The Shadow Knows	61
Problem 11	Phone Lines	63
Extension	Side Show	65

PROBLEM SET 12
Warm-up	Round About	67
Problem 12	I Never Promised You . . .	69
Extension	Equal Rights	71

PROBLEM SET 13
Warm-up	Crow's Feet	73
Problem 13	Looney Tunes	75
Extension	All Wet	77

PROBLEM SET 14
Warm-up	Diamond in the Round	79
Problem 14	A River Too Wide	81
Extension	Radials Please	83

PROBLEM SET 15
Warm-up	Sergeant's Walk	85
Problem 15	Walk Straight	87
Extension	Treasure Hunt	89

PROBLEM SET 16
Warm-up	One Up	91
Problem 16	Cannonball Run	93
Extension	Making Arrangements	95

PROBLEM SET 17
Warm-up	Square Units	97
Problem 17	A Perfect Fit	99
Extension	Crayona's Canvas	101

PROBLEM SET 18
Warm-up	Tightening Your Belts	103
Problem 18	Head Start	105
Extension	High Rollers	107

PROBLEM SET 19
Warm-up	At Rope's End	109
Problem 19	Partial Eclipse	111
Extension	In the Shade	113

PROBLEM SET 20
Warm-up	Semi-square	115
Problem 20	Seemore's Symbol	117
Extension	Integrally Yours	119

PROBLEM SET 21
Warm-up	Turn Down	121
Problem 21	Triangle Park	123
Extension	Writer's Rectangle	125

PROBLEM SET 22
Warm-up	Smalley's Folly	127
Problem 22	Four Square and Eight	129
Extension	Reflect on This	131

PROBLEM SET 23
Warm-up	Double or Nothing	133
Problem 23	Multi-pi	135
Extension	Fill It to the Rim	137

PROBLEM SET 24
Warm-up	Two for One	139
Problem 24	Revolutionary Figures	141
Extension	Drill Bit	143

PROBLEM SET 25
Warm-up	Got You Cornered	145
Problem 25	Super Match	147
Extension	Whatever's Right	149

PREFACE

No classroom problem-solving effort is possible without good problems. And good problems can't often be created in the midst of a classroom situation. It takes time and planning to create problems that allow for different solution strategies; that can be varied and extended; that provide a good chance for success; that have understandable math; and that ask students to apply what they know to new situations. Good geometry problems are hard to come by.

With this book I have tried to give a set of problems that supplement the standard first-year geometry course. I have tried to cover all the major topics in the course. I must confess that I have a particular liking for the Pythagorean Theorem and have included many problems that use it somewhere in their solutions. Since most students have seen this property before entering geometry, I have assumed a familiarity with the theorem even though some geometry courses don't get to it until later in the year. I've also scattered problems including area and volume concepts throughout the later portions of this book.

Often, students seem to make their way through math courses without ever getting a handle on problem solving. A mathematics supervisor by the name of Joe Dodson from the Winston-Salem/Forsyth County Schools in North Carolina published the following tongue-in-cheek guide to problem solving in the *North Carolina State Math Newsletter*. It presents a problem-solving approach that too many students take.

A Student's Guide to Problem Solving

Rule 1: If at all possible, avoid reading the problem. Reading the problem only consumes time and causes confusion.

Rule 2: Extract the numbers from the problem in the order in which they appear. Be on the watch for numbers written in words.

Rule 3: If Rule 2 yields three or more numbers, the best bet for getting the answer is adding them together.

Rule 4: If there are only two numbers which are approximately the same size, then subtraction should give the best results.

Rule 5: If there are only two numbers in the problem and one is much smaller than the other, then divide if it goes evenly—otherwise, multiply.

Rule 6: If the problem seems like it calls for a formula, pick a formula that has enough letters to use all the numbers given in the problem.

Rule 7: If the Rules 1-6 don't seem to work, make one last desperate attempt. Take the set of numbers found by Rule 2 and perform about two pages of random operations using these numbers. You should circle about five or six answers on each page just in case one of them happens to be the answer. You might get some partial credit for trying hard.

Rule 8: Never, never spend too much time solving problems. This set of rules will get you through even the longest assignment in no more than ten minutes with very little thinking.

With help, students can improve their problem-solving skills. In my classes, I make a special effort to set aside time specifically for problem solving. My students have non-routine problems to solve each week and keep a special section in their notebooks for all those problems. I have them keep all their work and any notes they take from class discussions. I also ask them to keep notes on any special techniques we talk about in class, such as finite differences. (I suggest they keep copies of the finite differences charts given in the back of this book and the Guide to Problem Solving chart, too.)

Students must turn in their work on problems at the end of each week. They aren't penalized if they can't solve the problems, since I am aiming for success. They are penalized, however, if they don't keep their notebooks together. (We do work through all the problems in class.)

Solving problems isn't always easy; sometimes it is hard work. But we can help students learn to make solving problems easier; to make it a task that is both satisfying and rewarding. We can show students that the pleasure in reaching a successful solution generally makes the effort worthwhile.

R. S.
July 1984

INTRODUCTION

Geometry Problems: One Step Beyond is a collection of 25 sets of problems for students in grades 10 through 12. Each set contains three related problems, including one main problem, a warm-up problem that eases students into the work of the main problem, and an extension problem that takes students into more difficult but related concepts.

Accompanying this problem book are 25 posters that display each of the main problems. These posters are spiral-bound in a calendar-style format with colorful cartoon illustrations. You may use these posters alone or with the corresponding worksheets in this book. The *Geometry Problems* book itself may be used independently of the posters.

About the Book

Each of the 75 problems—warm-ups, main problems, and extensions—appears in this book on a separate reproducible worksheet page, with space allowed for students to work out their solutions. You may duplicate the pages and distribute them as in-class assignments or as homework.

On the back of each student worksheet page is a discussion of the problem for the teacher, including the answer to the problem, clues to assist students in completing the problem, a solution (sometimes more than one) that shows where the answer came from, and teaching suggestions that warn about points of confusion, indicate ways to get started and follow-up, and provide interesting background for the problems.

Using the Problems

Problems are arranged to follow the standard tenth-grade geometry curriculum, starting with logical thinking, then properties of triangles, and leading into work with circles, area, and volume. There is no required order for using the 25 sets of problems, however; simply choose those sets that best suit your students' learning situation. Within each set of problems, it is best to assign the warm-up problem first, then the main problem, and finally the extension problem.

These problems are not the sort that can necessarily be solved in one sitting. You may spend as much as a week or more on a single problem, giving hints at key points along the way, letting students discover what they can and talking over what they've learned.

The problems in this book are intended to help you concentrate on the development of students' problem-solving skills. In these problems, students will be using mathematical knowledge they already have, but in new and different ways.

There is no one right way to approach a problem. Different problems will require different strategies; often a problem can be solved in more than one way. With geometry problems, the ability to draw a picture describing the problem and to apply the given information to that picture is essential.

You may find the chart on the next page a helpful tool for reminding your students of steps they should follow when they attempt a problem. You may wish to give a copy to each student. A poster form of the chart is available from Dale Seymour Publications.

A GUIDE TO PROBLEM SOLVING

To understand a problem, try these suggestions.	• Read the problem carefully. • Decide what you're looking for. • Find the important information.
To develop your plan, use some of these ideas.	• Guess and check. • Draw a picture. • Look for a pattern. • Make a model. • Act it out. • Use easier numbers. • Write a number sentence. • Make an organized list. • Make a table or chart. • Use logic. • Work backwards.
To check your work, follow these steps.	• Make sure you used *all* the important information. • Check any arithmetic you may have done. • Decide if your answer makes sense. • Write your answer in a complete sentence.

This chart is adapted from *Teaching Problem Solving: What, Why and How* by Randall Charles and Frank Lester, ©1982, Dale Seymour Publications.

NAME THAT LINE

If AG is different from GA, how many different names can you give this line?

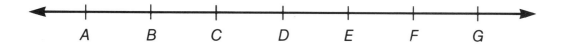

Discussion for

NAME THAT LINE

Answer: 42

Clues:

1. How many names have *A* as the first letter?
2. How many different choices are there for the first letter?
3. If you have 3 choices for the first task, and 2 choices for the second, in how many different ways could you accomplish the two tasks? (Think of a simpler problem.)

Solution:

Since it was given that $\overleftrightarrow{AG}$ differs from $\overleftrightarrow{GA}$, there are 7 different choices for the first letter of the name of the line. One of the seven letters has been chosen for the first letter in the name, so there are 6 different choices left for the second letter. So, there are 7 · 6, or 42, names. The names are:

$\overleftrightarrow{AB}$ $\overleftrightarrow{BA}$ $\overleftrightarrow{CA}$ $\overleftrightarrow{DA}$ $\overleftrightarrow{EA}$ $\overleftrightarrow{FA}$ $\overleftrightarrow{GA}$
$\overleftrightarrow{AC}$ $\overleftrightarrow{BC}$ $\overleftrightarrow{CB}$ $\overleftrightarrow{DB}$ $\overleftrightarrow{EB}$ $\overleftrightarrow{FB}$ $\overleftrightarrow{GB}$
$\overleftrightarrow{AD}$ $\overleftrightarrow{BD}$ $\overleftrightarrow{CD}$ $\overleftrightarrow{DC}$ $\overleftrightarrow{EC}$ $\overleftrightarrow{FC}$ $\overleftrightarrow{GC}$
$\overleftrightarrow{AE}$ $\overleftrightarrow{BE}$ $\overleftrightarrow{CE}$ $\overleftrightarrow{DE}$ $\overleftrightarrow{ED}$ $\overleftrightarrow{FD}$ $\overleftrightarrow{GD}$
$\overleftrightarrow{AF}$ $\overleftrightarrow{BF}$ $\overleftrightarrow{CF}$ $\overleftrightarrow{DF}$ $\overleftrightarrow{EF}$ $\overleftrightarrow{FE}$ $\overleftrightarrow{GE}$
$\overleftrightarrow{AG}$ $\overleftrightarrow{BG}$ $\overleftrightarrow{CG}$ $\overleftrightarrow{DG}$ $\overleftrightarrow{EG}$ $\overleftrightarrow{FG}$ $\overleftrightarrow{GF}$

Teaching Suggestions:

Some students might begin this problem by making a tree diagram to organize their thinking.

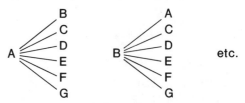

etc.

Clue 3 suggests a simpler, related problem that may help students recognize a solution plan. Once they see a multiplication pattern, encourage them to use this approach for solving the problem.

To help students appreciate the power of the multiplication method (over the listing method), you may wish to suggest a more complex problem for which counting every possibility is very tedious.

STAIR STEPS

What is the least number of additional walls needed to complete a stair-step pattern if you already have 350 walls?

4 walls
1 square per side

10 walls
2 squares per side

18 walls
3 squares per side

GEOMETRY PROBLEM 1

Discussion for
STAIR STEPS

Answer: 28 additional walls

Clues:

1. How many walls are needed to complete a stair pattern that has 4 squares per side? 5 squares per side? n squares per side?
2. Look for a formula relating the number of squares per side to the total number of walls.
3. The figure with 4 squares per side requires 28 walls.

Solution:

There are a variety of approaches to this problem. All of them depend on finding a general formula that relates the number of squares on a side to the total number of walls. For the purposes of discussion, let n stand for the number of squares per side and let T stand for the total number of walls.

EXAMINING FACTORS

Factor the values of T.

$4 = [1 \times 4]$ or 2×2
$10 = 1 \times 10$ or $[2 \times 5]$
$18 = 1 \times 18$ or 2×9 or $[3 \times 6]$ or $2 \times 3 \times 3$
$28 = 1 \times 28$ or 2×14 or $[4 \times 7]$ or $2 \times 2 \times 7$

Notice that each value of T has a pair of factors for which

- the factors differ by 3, and
- one of the factors is the corresponding value of n.

In other words, there is always a factorization for T that can be described by the following equation.

$$T = n(n + 3)$$

ADDING VERTICAL AND HORIZONTAL WALLS

One way to count the total number of walls for a given stair pattern is to count the number of vertical walls, count the number of horizontal walls, and add.

squares per side	1	2	3	4
vertical walls	2	5	9	14
horizontal walls	2	5	9	14

A close study of the sequence of vertical walls reveals a pattern. For n squares on a side, the number of vertical walls is $(1 + 2 + 3 + \cdots + n) + n$. The sum of the first n integers is $[n(n + 1)]/2$, making the number of vertical walls $([n(n + 1)]/2) + n$. The number of horizontal walls is the same value. Using this result, you can find a value for T.

total number walls = no. vertical + no. horizontal

$$T = \left(\frac{n(n+1)}{2} + n\right) + \left(\frac{n(n+1)}{2} + n\right)$$

$$= 2\left(\frac{n(n+1)}{2} + n\right)$$

$$= n(n+1) + 2n$$

$$= n^2 + 3n$$

$$= n(n+3)$$

USING FINITE DIFFERENCES

To find the general term of the sequence (T_n), first look at differences of successive terms in the given sequence.

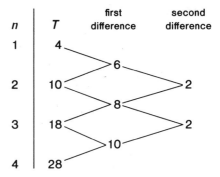

Since a common difference arises after two sets of subtractions, according to the method of finite differences the general term is a quadratic of the form $ax^2 + bx + c$.

x	$ax^2 + bx + c$	first difference	second difference
1	$a + b + c$		
2	$4a + 2b + c$	$3a + b$	$2a$
3	$9a + 3b + c$	$5a + b$	$2a$
4	$16a + 4b + c$	$7a + b$	

To find an expression for the general term, solve the following system of equations.

$$2a = 2$$
$$3a + b = 6$$
$$a + b + c = 4$$

You will find that $a = 1$, $b = 3$, and $c = 0$. This gives $T = (1)n^2 + (3)n + (0)$, or $n^2 + 3n$.

Once the formula relating n and T is found, it is possible to determine values of n for which the value of T is close to 350 by making a quick guess-and-check. For $n = 17$, $T = 340$. For $n = 18$, $T = 378$. Since the problem asks for *additional* walls needed, the answer is $378 - 350$, or 28 walls.

CONTINUED ON PAGE 151

REGIONAL PARTS

Each vertex of a triangle is joined by straight lines to 6 points on the opposite side of the triangle. No three of the joining lines pass through the same point. Into how many regions do these 18 lines divide the interior of the triangle?

Discussion for
REGIONAL PARTS

Answer: 127 regions

Clues:

1. Start with a simpler problem.
2. If there are 4 points on a side, there are 61 regions.
3. The answer is *not* 122.
4. Make a table of values starting with 1 point per side, then 2 points per side, and so on.

Solution:

Completing a diagram for this problem in which no three lines pass through the same point is an extremely difficult task. Also, the process of counting the triangles is next to impossible unless the diagram is very large. A systematic analysis of the construction of the diagram avoids the pitfalls of the counting process.

Consider triangle *ABC* where lines connect vertex *C* to 6 distinct points on side $\overline{AB}$. These lines separate triangle *ABC* into 7 small triangles.

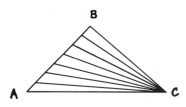

Now, draw one line from *B* to a point on $\overline{AC}$. This line consists of seven segments and each of these segments separates one of the small triangles into two pieces, giving a total of 14 regions. Therefore, drawing this line has increased by 7 the number of regions into which triangle *ABC* is separated.

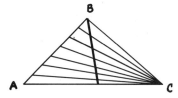

A second line drawn from *B* to $\overline{AC}$ increases the number of regions by 7, for a total of 21 regions.

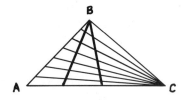

EXTENSION FOR PROBLEM 1

As additional lines are drawn from *B* to $\overline{AC}$, the total number of parts increases by 7 each time. Thus, at the end of the process, triangle *ABC* is separated into 7 + 6(7), or 7^2, regions—49 regions.

Each line segment drawn from vertex *A* to $\overline{BC}$ intersects all 12 segments previously drawn from vertex *B* to $\overline{AC}$ and from *C* to $\overline{AB}$. The points of intersection are all distinct since no line segments are concurrent. Thus, each segment emanating from vertex *A* is divided into 13 smaller segments and, consequently, each line segment from *A* increases the total number of regions by 13. It follows that the total number of regions is 49 + 6(13), or 127.

Teaching Suggestions:

Help students express a general formula for this problem. Let *n* stand for the number of points on each side. In the case of the given problem, *n* is 6. The argument given in the solution gives $(6 + 1)^2 + 6(2 \cdot 6 + 1)$, or $(n + 1)^2 + n(2n + 1)$, which simplifies to $3n^2 + 3n + 1$.

Although the complexity of the diagrams makes counting regions difficult, the problem can be solved by counting and using finite differences. By drawing pictures and counting, the following table can be obtained.

points per side	regions
1	7
2	19
3	37
4	61

A pattern begins to appear at this point and can be used to continue the chart. Examining successive subtractions reveals a common difference of six after the second set of subtractions.

points per side	regions		
1	7		
		12	
2	19		6
		18	
3	37		6
		24	
4	61		6
		30	
5	91		6
		36	
6	127		

TYPE CAST

The figure below consists of many different triangles. There are five different sizes of right triangle, three different sizes of equilateral triangle, and three different sizes of isosceles triangle (*not* equilateral).

Correctly sketch each different type of triangle and find how many of each type are in the figure. (For each sketch, show the lines in the triangle as well.)

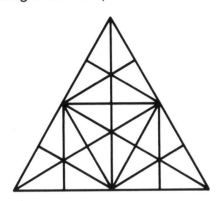

Discussion for
TYPE CAST

Answer: See Solution.

Clues:

1. Start with the smallest triangle. What kind of triangle is it and how many times does it appear?
2. Count the smallest triangles systematically. Notice that the entire figure can be separated into four identical equilateral triangles. Find the number of small triangles in one equilateral triangle and multiply by four.
3. Look for equilateral triangles.
4. Make a chart to help you keep track of the triangles you count.

Solution:

EQUILATERAL TRIANGLES

The equilateral triangles are probably easiest to identify. There are 6 small equilateral triangles (identified by A) which together form a hexagon inside the figure. There are 4 medium-sized equilateral triangles (B) that partition the figure. There is one large equilateral triangle (C).

RIGHT TRIANGLES

The smallest triangles in the figure (D) are right triangles and are all congruent to one another. Systematic counting reveals 24 in all.

The medium-sized equilateral triangles previously identified each contain 6 congruent right triangles (E). Since there are 4 medium-sized equilateral triangles, there are 4 × 6 of these right triangles.

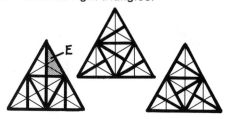

Each triangle (F) in the next set of right triangles is made up of four small triangles.

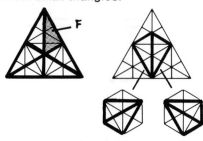

There are 6 of the second largest right triangles (G).

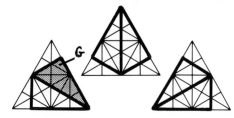

There are 2 large right triangles (H).

ISOSCELES TRIANGLES

The isosceles triangles are shown in the following diagrams.

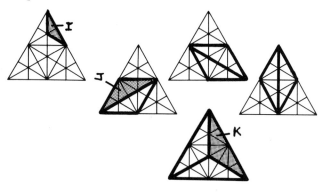

There are 12 of the small I triangles, 6 of the J triangles, and 3 of the K triangles.

Teaching Suggestions:

For classes having trouble discovering the different shapes, I pass out copies of the shapes and have students complete the problem by counting the shapes' appearances in the given figure.

You can use this problem later in the year, after students have studied 30-60 right triangles. You will also find that calculating the areas of the shapes in the figure proves to be an interesting exercise.

WARM-UP FOR PROBLEM 2

PENTAGON SEARCH

If every vertex of a regular pentagon is connected to every other vertex, how many triangles are formed?

Discussion for
PENTAGON SEARCH

Answer: 35 triangles

Clues:

1. Label each vertex of the pentagon (A_1, A_2, A_3, A_4, and A_5). Label each of the five interior points of intersection (B_1, B_2, B_3, B_4, and B_5).
2. How many triangles have only A-points as vertices? Can you name them?
3. How many triangles have only two consecutive A-points as vertices? Two non-consecutive A-points?

Solution:

The figure described by this problem is as follows.

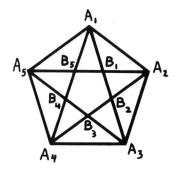

The vertices of the pentagon are A_1, A_2, A_3, A_4, and A_5. For convenience, they will be called A-points. The five interior points of intersection are B_1, B_2, B_3, B_4, and B_5.

One way to solve the problem is by classifying the different kinds of triangles that are formed and counting the triangles in each group.

TRIANGLES WITH 3 A-POINTS

There are 5 different A-points. According to combinatorics, 5 different points taken 3 at a time gives a total of 10 triangles—$C(5, 3)$. A list of the triangles is:

$A_1A_2A_3$ $A_1A_3A_4$ $A_1A_4A_5$
$A_1A_2A_4$ $A_1A_3A_5$
$A_1A_2A_5$ $A_2A_4A_5$

$A_2A_3A_4$
$A_2A_3A_5$
$A_3A_4A_5$

TRIANGLES WITH 2 CONSECUTIVE A-POINTS AND 1 INTERIOR POINT

First look at points A_1 and A_2. They can be connected with B_1, B_2, or B_5 to make 3 different triangles. There are 5 different pairs of A-points and each pair can be used to make 3 different triangles. In other words, there are 15 different triangles (5 × 3) with 2 consecutive A-points.

TRIANGLES WITH 2 NON-CONSECUTIVE A-POINTS AND 1 INTERIOR POINT

There are 5 different pairs of non-consecutive A-points: A_1A_3, A_2A_4, A_3A_5, A_4A_1, and A_5A_2. Each of these pairs can be connected with only one interior point, giving a total of 5 different triangles.

TRIANGLES WITH 1 A-POINT AND 2 INTERIOR POINTS

Look at point A_1. $A_1B_1B_5$ is the only triangle that can be constructed with 2 interior points and A_1. The same situation holds true for all 5 A-points, so there are 5 triangles in this category.

The total number of triangles is 10 + 15 + 5 + 5, or 35.

Teaching Suggestions:

You will find that most of your students have not been introduced to combinatorics, but that they will accept the results if you introduce the use of Pascal's triangle. I normally introduce this technique as early in the school year as possible so that the students have one more method to use when attacking problems.

I usually let students develop patterns in the triangle themselves. I start with the first three rows and then ask for student input. After the first four rows are filled in, most students are able to complete more rows by themselves.

Most of my students have had some work with sets and subsets, and I use this experience to develop their ability to use Pascal's triangle. For example, I relate the fifth row of Pascal's triangle to a set consisting of 5 elements by asking how many different subsets can be made having 5, 4, 3, 2, 1, and 0 elements, respectively. This exercise also reminds students that the empty set is a subset of any set.

Most students will try to do this problem by counting the triangles. They soon discover that they need to organize their work. The clues are designed to help students who have difficulty figuring out a logical approach to the problem.

GEOMETRY PROBLEM 2

TWO'S COMPANY

Suppose every vertex of a regular octagon is connected to every other vertex. How many triangles are formed that contain only *two* points of the octagon as vertices?

Discussion for
TWO'S COMPANY

Answer: 280 triangles

Clues:

1. Draw a diagram and label it systematically. Label the vertices of the octagon V_1 through V_8. Use the same letter to label interior points that are equidistant from the center of the figure.
2. How many triangles have V_1 and V_2 as vertices?
3. Find all the triangles that are formed from two consecutive points of the octagon.
4. Five triangles are formed with V_1, V_2, and a point on diagonal $\overline{V_1V_3}$. Can you name them?
5. How many triangles are formed with V_1, V_2, and a point on $\overline{V_1V_4}$?
6. The answer to Clue 2 is 15.
7. How many triangles have V_1 and V_3 as vertices? V_1 and V_4? V_1 and V_5? (Do *not* include triangles that have been counted earlier.)

Solution:

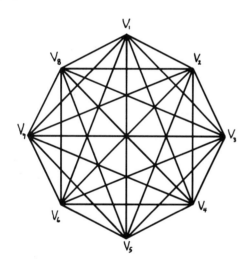

Four distinct types of triangles are formed—those including three vertices of the octagon, those including only two vertices of the octagon, those with only one, and those with none. This problem asks about triangles that include only two vertex points of the octagon—only two V-points.

There are four different types of triangles formed that have two vertices of the octagon. Patterns and organized listing will lead to the solution.

Teaching Suggestions:

When I first tried this problem with my students, I asked them to find *all* of the triangles formed. Since there are 632 triangles in the figure, this proved to be an arduous task. Now I present a scaled-down version of my question depending on the experience and tenacity of the given class of students. With my less experienced students, for example, I will ask how many triangles can be formed by using three vertices of the octagon or I will ask how many triangles have no vertices of the octagon at all.

An accurate picture for this problem is essential. I normally create an overhead transparency of the figure and go through at least part of one case (say the V_1-V_2 case) in class.

For your reference, the 632 triangles in the figure can be categorized as follows.

  56 with 3 octagon vertices
  280 with 2 octagon vertices
  280 with 1 octagon vertex
  16 with 0 octagon vertices

See the detailed graphic solution on page 152.

EXTENSION FOR PROBLEM 2

ALPHABET SOUP

Six geometry students, Al, Betty, Chuck, Dot, Ed, and Flo, took a college entrance examination. Given the following clues, rank the students in order from highest score to lowest score.

- Al and Betty had the same score.
- Al's score was higher than Chuck's.
- Chuck scored higher than Dot.
- Ed's score was lower than Al's, but higher than Dot's.
- Ed's score was lower than Chuck's.
- Betty's score was lower than Flo's.

Discussion for
ALPHABET SOUP

Answer: Flo, Al and Betty, Chuck, Ed, and Dot.

Clues:

1. Put each person's name on a separate slip of paper.
2. Assume they scored from highest to lowest alphabetically, by name, and then rearrange the slips clue by clue.
3. Omitting *B*, the first letters of each name placed in correct order together spell out an English word.

Solution:

Let *A*, *B*, *C*, *D*, *E*, and *F* stand for Al's, Betty's, Chuck's, Dot's, Ed's, and Flo's scores, respectively. Start by assuming their scores are arranged "alphabetically." Then rearrange the scores according to the information in the clues.

START:	*A*	*B*	*C*	*D*	*E*	*F*
AFTER CLUE 1:	*A*	*B*	*C*	*D*	*E*	*F*
AFTER CLUE 2:	*A*	*B*	*C*	*D*	*E*	*F*
AFTER CLUE 3:	*A*	*B*	*C*	*D*	*E*	*F*
AFTER CLUE 4:	*A*	*B*	*C*	*E*	*D*	*F*
AFTER CLUE 5:	*A*	*B*	*C*	*E*	*D*	*F*
AFTER CLUE 6:	*F*	*A*	*B*	*C*	*E*	*D*

Teaching Suggestions:

Most students do well on this type of problem with little help from the teacher. Some may need help explaining their solutions.

WARM-UP FOR PROBLEM 3

RAIN OR SHINE

The weather during the Smedleys' vacation was strange.

- It was cloudy on 13 different days, but it was never cloudy for an entire day.
- Cloudy mornings were followed by clear afternoons.
- Cloudy afternoons were preceded by clear mornings.
- There were 11 clear mornings and 12 clear afternoons in all.

How long was the vacation?

Discussion for

RAIN OR SHINE

Answer: 18 days

Clues:

1. Were there any days with both clear mornings and clear afternoons?
2. How many different kinds of days were there?
3. Try to write equations to describe the information.

Solution:

There were three different kinds of days during the Smedleys' vacation.

a. Days with cloudy mornings and clear afternoons.
b. Days with clear mornings and cloudy afternoons.
c. Days with both clear mornings and clear afternoons.

Let a, b, and c stand for the number of days in each category as listed above, respectively.

The problem gives three pieces of numerical information.

It was cloudy on 13 different days.
$$a + b = 13$$
There were 11 clear mornings.
$$b + c = 11$$
There were 12 clear afternoons.
$$a + c = 12$$

As shown, the information gives three equations in three unknowns that can be solved in a number of different ways.

$$a + b = 13$$
$$b + c = 11$$
$$a - c = 2$$

$$a + c = 12 \qquad \underline{a + c = 12}$$
$$2a = 14$$
$$a = 7$$

$$a - c = 2$$
$$7 - c = 2$$
$$c = 5$$
$$a + b = 13$$
$$7 + b = 13$$
$$b = 6$$

The total number of days on the vacation is $a + b + c$, or $7 + 6 + 5$, or 18.

Teaching Suggestions:

Most of my students try to solve the problem by simply adding up the numbers, arriving at an answer of 36 days. Other students try to use Venn diagrams, finding the approach unworkable. Once students look a little closer and realize there are only three different kinds of days on the vacation, they come up with the equations on their own and have no trouble solving them.

GEOMETRY PROBLEM 3

PROBLEM CHILD

Can the following conditional statements be arranged in logical order? If so, what theorem do they prove?

- If I can't understand a problem, I get confused while studying it.
- If a problem doesn't follow a pattern, then I can't understand it.
- If I don't get confused while studying a problem, I can understand it.
- If I get confused while studying a problem, then it is giving me trouble.
- If a problem is giving me trouble, then it is too hard.

Discussion for
PROBLEM CHILD

Answer: If this problem does not follow a pattern, then the problem is too hard.

Clues:

1. Rewrite all the statements in if-then form.
2. If A, then B. If B, then C. If A, then _____.
3. Of the inverse, converse, and contrapositive of a statement, which is logically equivalent to the original statement?
4. Two of the five statements are logically equivalent. Which two and why?

Solution:

Symbolize the parts of the conditional statements as follows.

$U = $ I understand a problem
$C = $ I get confused by studying a problem
$P = $ The problem follows a pattern
$T = $ The problem gives me trouble
$H = $ The problem is too hard

The conditional statements can be symbolized in the following way.

Theorem 1: If not U, then C.
Theorem 2: If not P, then not U.
Theorem 3: If not C, then U.
Theorem 4: If C, then T.
Theorem 5: If T, then H.

Theorems 1 and 3 are logically equivalent; they are contrapositives of one another. Only one of these two statements is needed to arrange a logical syllogism. In this particular case, Theorem 1 will be used.

The statements should be arranged so that the conclusion of one becomes the hypothesis of the next.

Theorem 2: If not P, then not U.
Theorem 1: If not U, then C.
Theorem 4: If C, then T.
Theorem 5: If T, then H.

This sequence of statements, if assumed, can be used to prove *If not P, then H*. *If not P* is the hypothesis of the very first statement; *then H* is the conclusion of the very last statement.

Teaching Suggestions:

If each of the statements is taken to be a proved theorem, then a two-column proof can be set up as follows.

GIVEN: This problem does not follow a pattern. Theorems 1, 2, 4, and 5.
TO PROVE: This problem is too hard

STATEMENTS	REASONS
1. This problem does not follow a pattern.	1. Given
2. I can't understand this problem	2. Theorem 2
3. I get confused while studying this problem.	3. Theorem 1
4. This problem gives me trouble.	4. Theorem 4
5. This problem is too hard.	5. Theorem 5

EXTENSION FOR PROBLEM 3

FRUIT STAND

A man selling fruit has only three weights, but with them he can weigh any whole number of kilograms from 1 kg up to 13 kg inclusive on his balance. What weights does he have?

Discussion for

FRUIT STAND

Answer: 1 kg, 3 kg, and 9 kg

Clues:

1. The balance used is a pan balance.
2. Weights can be put in either pan.
3. The difference between the masses of the weights in the two pans represents the mass of the fruit. (For example, if an apple and a weight of mass a kg are in one pan and weights with masses of b kg and c kg are in the other pan, then the mass of the apple is $[(b + c) - a]$ kg.)

Solution:

For the purpose of getting started, assume that the mass of each weight is less than 13 kg and that the three weights together have a total mass of 13 kg. To maximize the number of possible sums and differences among the weights, also assume that no two weights have the same mass. Under these assumptions, the possible choices of weights are as follows.

1 kg, 2 kg, 10 kg	2 kg, 3 kg, 8 kg
1 kg, 3 kg, 9 kg	2 kg, 4 kg, 7 kg
1 kg, 4 kg, 8 kg	2 kg, 5 kg, 6 kg
1 kg, 5 kg, 7 kg	3 kg, 4 kg, 6 kg

Theoretically, there are 13 possible different combinations of these weights that can be made using additions and subtractions. If the weights have masses of a kg, b kg, and c kg with $a < b < c$, then the sums and differences are:

a	$a + b$	$b - a$	$(b + c) - a$
b	$a + c$	$c - a$	$(a + c) - b$
c	$b + c$	$c - b$	$(a + b) - c$
	$a + b + c$		

By systematically checking the sums and differences of the possible choices of weights, you will find that only one combination of those given will yield all the different whole numbers from 1 through 13. That combination is 1kg, 3kg, and 9 kg. The numbers can be created as follows.

1:	1 kg
2:	3 kg − 1 kg
3:	3 kg
4:	1 kg + 3 kg
5:	9 kg − (1 kg + 3 kg)
6:	9 kg − 3 kg
7:	(1 kg + 9 kg) − 3 kg
8:	9 kg − 1 kg
9:	9 kg
10:	1 kg + 9 kg
11:	(3 kg + 9 kg) − 1 kg
12:	3 kg + 9 kg
13:	1 kg + 3 kg + 9 kg

Teaching Suggestions:

The key to this problem is understanding that the man can put any combination of weights in *either* pan and that the difference between the weights in the two pans represents the mass of the fruit being weighed. A few concrete examples are very helpful in getting the point across.

The values of the weights in this solution are in powers of three—3^0, 3^1, and 3^2. You may want to make the connection for students and point out that often problems looking for combinations in the way this problem does have solutions of this type.

You may also wish to try extending this problem to greater numbers of weights and greater values to weigh. For example, you may wish to consider the problem for 4 weights measuring from 1 kg up to and including 40 kg. You might also wish to restrict conditions of the given problem, allowing weighings up to only 11 kg, for example. And you might ask what would happen if you tried powers of 2 rather than powers of 3.

WARM-UP FOR PROBLEM 4

AFTERNOON AT THE PREVU

The Prevu Theater keeps a set of five counterweights to help stagehands move heavy props. The weights can balance exactly any load that is a multiple of 10 kg, from 10 kg up to a total of the five weights. Give the masses of weights which meet these conditions and allow the stagehands to counterbalance the maximum possible load.

Discussion for

AFTERNOON AT THE PREVU

Answer: 10 kg, 20 kg, 40 kg, 80 kg, and 160 kg

Clues:

1. When counterbalancing loads, the weights can be attached to only one end of the rope.
2. No two of the five weights have the same mass.
3. What is the mass of the lightest weight?
4. Try a simpler related problem first.
5. Any whole number of kilograms from 1 kg up to and including 7 kg can be weighed using just three weights.

Solution:

Let a, b, c, d, and e stand for the masses of the five weights. The possible weighings are given by the following list.

```
a        a+b      a+b+c    a+b+c+d
b        a+c      a+b+d    a+b+c+e
c        a+d      a+b+e    a+b+d+e
d        a+e      a+c+d    a+c+d+e
e        b+c      a+c+e    b+c+d+e
         b+d      a+d+e
         b+e      b+c+d
         c+d      b+c+e
         c+e      b+d+e
         d+e      c+d+e
a+b+c+d+e
```

The list shows that there are *at most* 31 different weighings possible because there are 31 different combinations of the five weights. If it is possible to choose values for the masses of the weights so that each combination of weights gives a *different* mass, then the maximum possible total mass for the weights is 310 kg.

Since only loads in multiples of 10 kg are considered, the mass of each weight is a multiple of 10 kg. The greater the mass of each weight, the greater the sum of their masses and, hence, the greater load they can counterbalance. In order to achieve the maximum possible mass then, the lightest weight must have a mass of 10 kg to counterbalance the lightest load.

The second lightest weight must have a greater mass than the 10 kg weight. The next multiple of 10 is 20. Weights of 10 kg and 20 kg allow the stagehands to counterbalance loads of 10 kg, 20 kg, and 30 kg.

The third weight must have a mass of 40 kg since the first two weights can measure only up to 30 kg. The addition of this 40 kg weight to the collection makes possible the counterbalancing of 40 kg, 50 kg, 60 kg, and 70 kg.

Continuing in this fashion gives a series of weights—no two having the same mass and every possible combination having a different mass—with masses of $10 \cdot 2^0$ kg, $10 \cdot 2^1$ kg, $10 \cdot 2^2$ kg, $10 \cdot 2^3$ kg, and $10 \cdot 2^4$ kg. These values total to 310 kg, the maximum possible load.

Teaching Suggestions:

Be sure students realize that loads can be counterbalanced only with additions of weights (*not* additions or subtractions as is the case with WARM-UP 4).

Once students realize they are looking for combinations of multiples of 10 up to 310, you may wish to simplify the problem by suggesting that they find whole number combinations for numbers from 1 to 31. Then you can suggest they simplify the problem further by looking at three whole numbers that will give combinations from 1 to 7. Working from the bottom up, trying to maximize the number of loads with each addition of a weight, will get students on the right track.

GEOMETRY PROBLEM 4

IN THE BALANCE

An apothecary has a set of five weights for use in the pans of his balance. By proper selection of weights, he is able to measure every multiple of 0.5 g, from 0.5 g up to a total of the five weights together. If the arrangement of weights is such that the apothecary can weigh the maximum possible amount, what are the five weights?

Discussion for

IN THE BALANCE

Answer: 0.5 g, 1.5 g, 4.5 g, 13.5 g, and 40.5 g

Clues:

1. Weights can be put in either pan.
2. Both addition and subtraction can be used to give possible amounts.
3. You may find the problem easier to work with if you first solve for units of 1 g and then multiply your answers by 0.5 (or divide by 2).
4. WARM-UP 4 is a simpler, related problem.

Solution:

Since whole numbers are easier to work with, first consider increments of 1 g starting with 1 g and going up to a total of five weights.

The lightest weight must have a mass of 1 g. A mass of 2 g can be achieved by adding another 1 g weight, by using a 2 g weight, or by finding the difference between the 1 g weight and a 3 g weight. The 3 g weight gives the additional benefit of allowing you to weigh amounts of 3 g and 4 g. Neither an extra 1 g weight nor a 2 g weight allow you to weigh amounts up to 4 g, so the second lightest weight must be a 3 g weight.

A third weight with a mass of 9 g allows you to measure all loads up to and including 13 g. Weights with masses less than 9 g will not allow you to go that high; weights greater than 9 g will not allow you to measure *all* the different loads.

The first three weights have masses of 3^0 g, 3^1 g, and 3^2 g, respectively. The emerging pattern suggests that the next weight should have a mass of 3^3 g and that the heaviest weight should have a mass of 3^4 g. That is, the five weights have masses of 1 g, 3 g, 9 g, 27 g, and 81 g, respectively.

To solve the given problem, multiply each value by 0.5, giving values of 0.5 g, 1.5 g, 4.5 g, 13.5 g, and 40.5 g and allowing you to weigh amounts from 0.5 g up to and including 60.5 g in increments of 0.5 g.

Teaching Suggestions:

You could borrow a pan balance and some gram weights from your science department to use in giving a physical demonstration of this problem (or a simpler related problem). Students are apt to find the solution method and the resulting numbers more convincing if they have actually seen how the solution works.

Be sure students realize that the weights can be put into both pans. In other words, amounts can be weighed through additions or subtractions.

WARM-UP PROBLEM 4 is an excellent starting point for getting students into this problem. It introduces them to the solution method without overwhelming them by the numbers and quantities of weights. The warm-up also sets students up for the numerical pattern that emerges.

EXTENSION FOR PROBLEM 4

CUBED SALAD

Tell whether or not a cube can be cut with a plane to produce the following figures.

- an isosceles triangle (not equilateral)
- an equilateral triangle
- a square
- a rhombus (*not* a square)
- a rectangle
- a regular pentagon
- a regular hexagon
- any polygon having more than six sides.

If the cut is possible, show the cut.

Discussion for

CUBED SALAD

Answer: 1, 2, 3, 4, 5, and 7 are possible. See solutions below for suggested cuts.

Clues:

1. To produce a triangle, how many faces of the cube must be cut?
2. How many faces does a cube have?
3. Try using cutting through just vertices of the cube, or just midpoints of sides, or both to create the figures.
4. Make the desired figures as large as possible.

Solution:

The following solutions are not unique. (Be prepared to accept different solutions as long as they can be justified).

ISOSCELES TRIANGLE: Any cut using one vertex of the cube for a vertex and cutting through an opposite face so that it slices off edges of equal lengths will make an isosceles triangle.

EQUILATERAL TRIANGLE: Cutting through any three vertices, no two of which are adjacent, will form an equilateral triangle since the diagonals of each face of the cube are congruent. Cutting off a corner so that the edge distances from the corner vertex to the vertices of the triangle are the same will form an equilateral triangle.

SQUARE: Slicing through the cube parallel to any face will form a square.

RHOMBUS: A plane passing through diagonally opposite vertices and bisecting the edges that it intersects will form a rhombus.

RECTANGLE: A slice going through any two diagonally opposite sides will form a rectangle. Slices parallel to that slice will also form rectangles.

PENTAGON: The figure shows one possible solution.

REGULAR HEXAGON: The figure shows how to form a regular hexagon. Each vertex is at the midpoint of an edge, creating congruent sides for the hexagon. Since each angle is formed in exactly the same way, the angles are congruent.

POLYGON WITH MORE THAN 6 SIDES: Since a cube has only six faces, the maximum number of sides a figure can have is six.

Teaching Suggestions:

This problem is very straightforward, but can be made much more difficult by asking students to find lengths of the sides of the figures they produce or to create the largest possible figure for each part of the problem.

Models are extremely helpful for visualizing the solutions to these problems. The best model to use is a cube that has only the edges (is not solid). Using straws or any other type of rod to represent the cuts, it is easy to show each of the resulting figures.

CONTINUED ON PAGE 151

WARM-UP FOR PROBLEM 5

LINE UP

How many different cubes can you make if each face of a given cube has a line connecting the center points of two opposite edges?

Discussion for

LINE UP

Answer: 8 cubes

Clues:

1. How many faces does a cube have?
2. Is it possible to go around the cube with one continuous line?
3. Is it possible to draw lines so that no one line touches any other?
4. How many lines can be drawn if there are four lines forming one continuous line? Three lines? Two lines?

Solution:

There are four different situations that can arise.

FOUR LINES FORMING ONE CONTINUOUS LINE (a square) AROUND THE CUBE. The other two lines can be parallel to one another (a) or they can be skew lines (b).

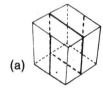
(a)
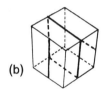
(b)

THREE LINES FORMING A CONTINUOUS LINE (three sides of a square). There can be two sets of three connected segments (c). The other two examples have one set of three connected segments, one with two lines connected (d) and the other with the remaining three lines not connected (e).

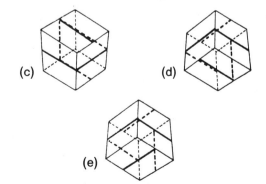

THREE PAIRS OF LINES, EACH PAIR FORMING TWO SIDES OF A SQUARE. Two such systems are possible (f and g); one is a mirror reflection of the other.

(f)
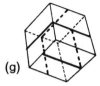
(g)

ALL SIX LINES ISOLATED (h).

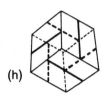
(h)

Teaching Suggestions:

Most students believe there are many more possibilities than actually exist. Having students draw each possibility is essential. They may have counted a particular possibility more than once and they may have missed others. A cardboard model with lines drawn on it will help convince them that one possibility may be the same as another upon rotation.

GEOMETRY PROBLEM 5

SURFACE PAINT

Each cube below is made up of smaller cubes, but the large cubes are *not* solid. They have tunnels through them.

- The first cube originally had 27 small cubes, but the tunnel removed 3 cubes.
- The second cube originally had 64 small cubes, but two straight tunnels, 4 cubes deep, removed some cubes.
- The third cube, which originally had 125 small cubes, has 3 straight tunnels, five cubes deep from face to face.

The outside surfaces of these cube constructions have been painted *including* inside the tunnels and on the bottom. For each construction, how many small cubes have paint on 4 faces? 3 faces? 2 faces? 1 face? 0 faces?

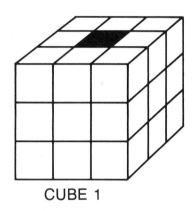
CUBE 1

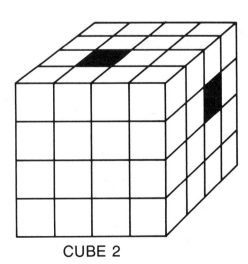
CUBE 2

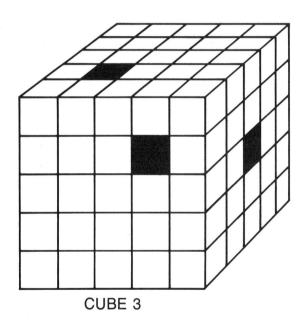
CUBE 3

Discussion for
SURFACE PAINT

Answer: See Solution.

Clues:

1. How many small cubes are you working with in each problem?
2. Other than the corner pieces, where are the pieces that have three painted faces located?
3. There are no cubes with 0 faces painted in either the 3-unit cube nor the 4-unit cube.
4. Draw a picture or make a model to help organize the counting.

Solution:

Shown below are layer diagrams which account for all of the cubes. The numbers in each square indicate the number of painted faces for the cube represented.

Adding up the numbers gives the results shown in the table below.

number of faces painted	3 × 3 × 3 cube	4 × 4 × 4 cube	5 × 5 × 5 cube
4	0 cubes	0 cubes	0 cubes
3	16 cubes	19 cubes	18 cubes
2	8 cubes	29 cubes	59 cubes
1	0 cubes	8 cubes	28 cubes
0	0 cubes	0 cubes	5 cubes

Teaching Suggestions:

Students will find that actually building models is the most satisfactory approach to solving this problem. When I did the problem, I started with the bottom layer and made sure of this layer before I built the next layer.

I have found that, in drawing out each layer, adding a broken line through a cube to signify that the top or bottom of that cube faces a tunnel is very helpful.

EXTENSION FOR PROBLEM 5

AT THE VERTEX

A quadrilateral is drawn in such a way that three of its sides are congruent, and the fourth side is congruent to both of the diagonals. Find the measures of the vertex angles in the quadrilateral.

Discussion for

AT THE VERTEX

Teaching Suggestions:

The figure is a trapezoid with AD parallel to BC. Avoid using the same variable for the alternate interior angles, since the key to the solution is to find two expressions in two variables. Some students may come up with a different set of equations. One of the equations could have been $y = 2x$, for example. Other students will solve this problem using only one variable.

Answer: Two adjacent angles are 108°. The other two adjacent angles are 72°. (See figure shown in solution.)

Clues:

1. Label the quadrilateral ABCD with base AD.
2. Classify triangles ABC and DCB.
3. Which triangles are congruent?
4. Are there any isosceles triangles? Which triangles?
5. Let x stand for the measure of the base angles in one set of isosceles triangles and y stand for the measure of the base angles in the other set.
6. Make up two equations using x and y.

Solution:

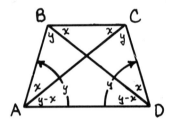

The figure is drawn so that $AB = BC = CD$ and $AD = AC = DB$ as given.

From this information, it follows that triangles ABC and BCD are both congruent (SSS) and isosceles. Let x stand for the measure of the base angles, ∠BAC and ∠BCA, ∠CBD and ∠CDB. Triangles ACD and DBA are also congruent (SSS) and isosceles. Let y stand for the measure of these base angles, ∠ACD and ∠ADC, ∠DBA and ∠DAB. It follows that the measure of ∠BDA is $y - x$.

The angle measures of triangle DBA total 180°

$$3y - x = 180$$

The angle measures of triangle ABC total 180.

$$y + 3x = 180$$

Solving these two equations simultaneously gives $x = 36$ and $y = 72$. Therefore, the base angles of quadrilateral ABCD measure 72° and the other two angles measure 108°.

WARM-UP FOR PROBLEM 6

STAR SUM

Find the sum of the measures of the five acute angles that make up this star.

Discussion for

STAR SUM

Answer: 180°

Clues:

1. Draw the pentagon in the interior of the star. Assume that the sides of the pentagon form straight lines with the sides of the star.
2. Label the points of the star and each interior angle of the pentagon.
3. Write equations that relate measures of the angles of the pentagon to measures of the angles of the star points.
4. Look for overlapping triangles. The sum of the degree measures of the interior angles of any convex pentagon is 540°.
5. Suppose the vertices of the star were points on a circle. What information could you surmise?

Solution:

Draw a pentagon by connecting the concave vertices of the star. Label the interior angles of the pentagon from 1 to 5 and label the points of the star from A to E.

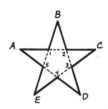

Notice that five different triangles are formed, each with one vertex a vertex of the pentagon and the other two vertices vertices of the star. Using the Angle-Sum Theorem you can write the following equations.

$m(\angle 1) + m(\angle C) + m(\angle E) = 180$

$m(\angle 2) + m(\angle A) + m(\angle D) = 180$

$m(\angle 3) + m(\angle B) + m(\angle E) = 180$

$m(\angle 4) + m(\angle A) + m(\angle C) = 180$

$m(\angle 5) + m(\angle B) + m(\angle D) = 180$

Add the equations together. Simplifying yields the following equation.

$[m(\angle 1) + m(\angle 2) + m(\angle 3) + m(\angle 4) + m(\angle 5)] +$
$2[m(\angle A) + m(\angle B) + m(\angle C) + m(\angle D) + m(\angle E)] = 5 \cdot 180$

But the sum of the angle measures of a convex pentagon is 540°.

Sum of angles $= (n - 2)180$ where n = no. angles
$= (5 - 2)180$
$= 540$

Substituting 540 in the equation gives the result.

$540 + 2[m(\angle A) + m(\angle B) + m(\angle C) + m(\angle D) + m(\angle E)] = 900$

$2[m(\angle A) + m(\angle B) + m(\angle C) + m(\angle D) + m(\angle E)] = 360$

$[m(\angle A) + m(\angle B) + m(\angle C) + m(\angle D) + m(\angle E)] = 180$

Teaching Suggestions:

Once students draw in the pentagon and recognize the appropriate triangles, they usually sail right through the problem. There are other approaches to the problem, such as looking at the angles of the pentagon as exterior angles to the triangles formed by the points of the star. You may wish to challenge students to find more than one solution.

My challenge to students is to extend the problem by increasing the number of points on the given star. I suggest that they consider only those stars with an odd number of points and that the stars are formed by connecting every other vertex.

This extension problem proves to be more difficult than most students anticipate because they need to deal with polygons other than triangles. For example, in the case of the seven-pointed star, students use overlapping quadrilaterals.

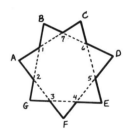

$m(\angle 1) + m(\angle C) + m(\angle G) + m(\angle E) = 360$
$m(\angle 2) + m(\angle A) + m(\angle D) + m(\angle F) = 360$
$m(\angle 3) + m(\angle B) + m(\angle E) + m(\angle G) = 360$
$m(\angle 4) + m(\angle A) + m(\angle C) + m(\angle F) = 360$
$m(\angle 5) + m(\angle B) + m(\angle D) + m(\angle G) = 360$
$m(\angle 6) + m(\angle A) + m(\angle C) + m(\angle E) = 360$
$m(\angle 7) + m(\angle B) + m(\angle D) + m(\angle F) = 360$

Simplifying gives a sum of 540° for the sum of the measures of the acute angles.

There are a number of ways to keep going with this problem. Students can try to find the answer for an n-pointed star where n is odd and the figure is formed by connecting every other vertex. They may also wish to generalize the solution. You might ask them to consider stars formed by connecting every third vertex.

GEOMETRY PROBLEM 6

HOW MANY SIDES?

How many sides does a polygon have if its smallest interior angle is 120° and each successive angle is 5° greater than its predecessor?

Discussion for
HOW MANY SIDES?

Answer: Either a 9-sided figure or a 16-sided figure. (In the 16-sided figure, one of the angles is 180°, making it, in fact, a 15-sided figure—a pentadecagon.)

Clues:

1. Represent each angle of the required figure by $120, 120 + 5, 120 + 10, \ldots, 120 + 5(n - 1)$.
2. Remember Gauss!
3. $5 + 10 + 15 + 20$ could be written as $5 \cdot 1 + 5 \cdot 2 + 5 \cdot 3 + 5 \cdot 4$ or $5(1 + 2 + 3 + 4)$.
4. Set up an equation involving the sum of the angles of the required n-gon and the sum of the angles of any polygon.

Solution:

The measures of the angles of the polygon are

$$120, 120 + 5, 120 + 5(2), 120 + 5(3), \ldots, 120 + 5(n - 1)$$

where n stands for the number of sides of the polygon. Adding gives the sum of the measures of the angles of the polygon.

$$120 + [120 + 5] + [120 + 5(2)] + \cdots + [120 + 5(n - 1)]$$
$$= 120n + 5[1 + 2 + 3 + \cdots + (n - 1)]$$
$$= 120n + 5\left[\frac{(n-1)(n)}{2}\right]$$
$$= \frac{5n^2 + 235n}{2}$$

The commonly used expression for the sum of the measures of the interior angles of an n-sided polygon is $180(n - 2)$. By equating the two expressions, you can find a value for n.

$$180(n - 2) = \frac{5n^2 + 235n}{2}$$
$$180n - 360 = \frac{5n^2 + 235n}{2}$$
$$360n - 720 = 5n^2 + 235n$$
$$0 = 5n^2 - 125n + 720$$
$$0 = n^2 - 25n + 144$$
$$0 = (n - 9)(n - 16)$$

So, n is either 9 or 16.

Teaching Suggestions:

Students need to be led through the derivation of the formula for the required n-gon. There are different ways to come up with a formula to find the sum of successive numbers, but I find the following one most effective with my students.

$$\text{Let } S = 1 + 2 + 3 + \cdots + n$$

Then, reverse the numbers to get

$$S = n + (n - 1) + (n - 2) + \cdots + 1$$

Adding these equations gives $2S = n(n + 1)$ since each column sums to $(n + 1)$ and there are n columns to sum. Dividing both sides by 2 gives

$$S = \frac{n(n - 1)}{2}$$

For the particular problem we need to add $(n - 1)$ multiples of five. The problem simplifies to adding successive numbers from 1 to $(n - 1)$ rather than from 1 to n. Using this technique results in a column sum of n and $(n - 1)$ columns to sum, and gives a total sum of $[(n)(n - 1)]/2$.

EXTENSION FOR PROBLEM 6

HALF A HEX

Given a regular hexagon and a point in its plane, draw a straight line through the given point to separate the hexagon into two congruent parts.

Discussion for

HALF A HEX

Answer: The straight line should pass through the center of the hexagon.

Clues:

1. Draw several different lines through the hexagon so they separate the hexagon into two congruent parts. (Don't worry about the given point.) Look for what these lines may have in common.
2. Look for concurrences with the lines you drew for Clue 1.
3. The point of concurrency of lines of symmetry of a figure is the center of symmetry of that figure.

Solution:

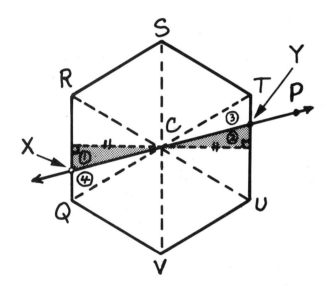

Let P represent the point given and let C represent the center of the regular hexagon.

Connecting vertices to the center of the hexagon forms six congruent triangles (by side-side-side). Two of these triangles contain the line $\overleftrightarrow{CP}$. Draw apothems of the hexagon in these triangles. These apothems separate the triangles into two congruent parts (hypotenuse-leg).

To show that $\overleftrightarrow{CP}$ separates the hexagon into two congruent parts, you need to show that triangle 1 is congruent to triangle 2 and that triangle 3 is congruent to triangle 4.

The central angles of triangles 1 and 2 are vertical and, therefore, congruent. Apothems of a hexagon are congruent, so the corresponding apothems of triangles 1 and 2 are congruent. Both triangles, 1 and 2, are right triangles. By angle-side-angle the two triangles are congruent.

Using a similar argument, you can prove that triangles 3 and 4 are congruent.

By "adding" the various congruent parts together, you can show that all corresponding parts of figures XRSTY and figure YUVQX are congruent and, consequently, the two figures are congruent.

Teaching Suggestions:

The solution here takes a traditional Euclidean approach to the problem. Paper folding gives a quick, intuitive solution to the problem. Some students may be interested in approaching the problem from a transformational standpoint.

In any case, the key to this problem is in realizing that all lines of symmetry for the regular hexagon go through the center of the figure. The clues are designed to help you lead students to that discovery. Essentially these clues ask students to look at a series of simpler related problems, find a common thread running through each of the solutions, and use it to solve the problem at hand.

Any figure with point symmetry has the same property as the one discovered here. You may wish to have students consider other regular polygons, finding which have point symmetry.

Congruent figures have equal areas. You may wish to restate this problem (or have students restate it) in terms of equal areas.

WARM-UP FOR PROBLEM 7

AS THE CROW FLIES

Norbert Finsterwald never gets things quite right. Take, for example, this map he drew of his neighborhood. Point out at least three mistakes Norbert made. (All distances are as the crow flies and the map is drawn roughly to scale.)

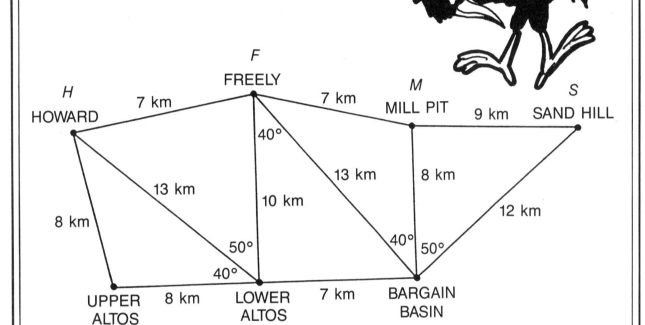

GEOMETRY PROBLEM 7

Discussion for

AS THE CROW FLIES

Answer:

Answers will vary. The most obvious contradictions are as follows.

- The distance from FREELY to MILL PIT should equal the distance from HOWARD to UPPER ALTOS according to the congruent triangles in Norbert's diagram, but he says the distances are not the same.
- The angle at FREELY, ∠LFB, should be congruent to the angle at LOWER ALTOS, ∠FLH according to the congruent triangles in Norbert's diagram, but he says they have different degree measures.
- The distance from FREELY to MILL PIT to SAND HILL should be greater than the distance from FREELY directly to SAND HILL which is about 17.69 km. But, according to Norbert, the distance is less.

Clues:

Please read the Teaching Suggestions before reading the Solution.

1. Assume parts of the diagram are correct and look for logical contradictions.
2. Look for congruent triangles and right triangles.
3. How are congruent triangles related?

Solution:

If you take a quick look at the UPPER ALTOS-LOWER ALTOS-HOWARD triangle, △ULH, and the MILL PIT-BARGAIN BASIN-FREELY triangle, △MBF, you will notice that by side-angle-side, they are congruent. Therefore, the distance from HOWARD to UPPER ALTOS must equal the distance from FREELY to MILL PIT. However, on Norbert's map, one distance is 8 km and the other is 7 km—a logical contradiction.

By side-side-side, the HOWARD-FREELY-LOWER ALTOS triangle, △HFL, is congruent to the BARGAIN BASIN-LOWER ALTOS-FREELY triangle, △BLF. Since corresponding parts of congruent triangles are congruent, the angle at FREELY, ∠LFB, should be congruent to the angle at LOWER ALTOS, ∠FLH. According to Norbert, however, one angle measures 40° and the other measures 50°.

A right angle is formed at BARGAIN BASIN, namely ∠FBS. Therefore, you can use the Pythagorean Theorem to find the direct distance from FREELY to SAND HILL.

$$FS^2 = FB^2 + BS^2$$
$$= 13^2 + 12^2$$
$$= 313$$
$$\approx 17.69$$

The shortest distance between two points is a straight line, so $FS < FM + MS$. But, according to Norbert, $FM + MS = 7 + 9 = 16$!

Teaching Suggestions:

When I use this problem with my students, I really don't expect them to go through this formal an approach. Using congruence relationships and corresponding parts of congruent triangles provides them with the desired contradictions; for example, △MBF ≅ △ULH by S.A.S. Examining angles opposite the 40° angle gives the lengths of 7 and 8 for the sides opposite, contradicting the corresponding parts theorem.

Students may find it easier to follow their work if they label each town with a letter. It is extremely important for them to keep track of where they have started in order to know where they are headed. Without a little help, some students are likely to go around in circles, not getting anywhere.

Reviewing congruence relationships will lead students along the right path to solving this problem. Norbert's major problem is with the HOWARD-UPPER ALTOS-LOWER ALTOS triangle and the FREELY-BARGAIN BASIN-MILL PIT triangle.

GEOMETRY PROBLEM 7

SIZE 'EM UP

Using the angle measures shown in the figures below, find the shortest and longest segments that are sides of a triangle.

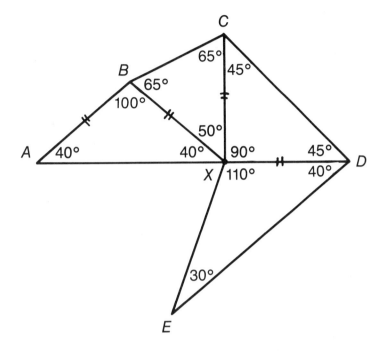

Discussion for

SIZE 'EM UP

Answer: $\overline{BC}$ is the shortest segment.
$\overline{DE}$ is the longest segment.

Clues:

1. Find the measures of all the angles.
2. Which triangles are isosceles?
3. The shortest segment is *not* the side opposite the smallest angle.
4. Use the Hinge Theorem which is described in the Solution.
5. $\overline{AX}$ is *not* the longest side.

Solution:

Applying the Angle Sum Theorem gives the remaining angle measures. Three different triangles have base angles congruent, so their corresponding sides are congruent. The following diagram shows the information.

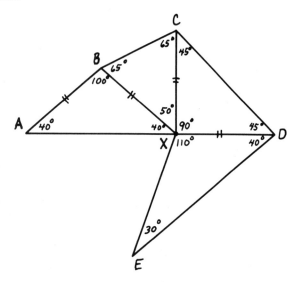

FINDING THE SHORTEST SEGMENT. If the measures of two angles of a triangle are unequal, then the measures of the sides opposite those angles are unequal in the same order. This theorem suggests that

- in triangle *ABX*, the shortest sides are $\overline{AB}$ and $\overline{BX}$;
- in triangle *BCX*, the shortest side is $\overline{BC}$, and $BC < BX$;
- in triangle *CDX*, the shortest sides are $\overline{CX}$ and $\overline{DX}$;
- in triangle *EDX*, the shortest side is $\overline{DX}$, and $DX < EX < DE$.

In other words, $BC < BX = AB = CX = DX$. So, the shortest segment is $\overline{BC}$.

FINDING THE LONGEST SEGMENT. Using the results from above, the candidates for longest segment are $\overline{AX}$, $\overline{CD}$, and $\overline{DE}$.

According to the Hinge Theorem, if two sides of one triangle are congruent to two sides of another triangle, but the measures of the included angles are unequal, then the measures of the third sides are unequal in the same order. Therefore, $CD < AX$.

To determine the relationship between $\overline{AX}$ and $\overline{DE}$, locate a point A' on $\overline{DE}$ so that the measure of angle DXA' is 100°. Since m($\angle DXA'$) < m($\angle DXE$), the point A' is located between D and E.

By angle-side-angle, triangle DXA' and triangle XBA are congruent. Therefore, $A'D = AX$. But, $A'D < DE$, so $AX < DE$ and $\overline{DE}$ is the longest segment.

Teaching Suggestions:

This problem gives a good introduction to the Hinge Theorem. I suggest presenting the problem two or three days prior to making a formal introduction of the Hinge Theorem. The more independent thinkers in your class, in particular, will be stimulated by such an approach.

EXTENSION FOR PROBLEM 7

LIGHTHOUSE

The captain of a ship at point A and sailing toward point B observes a lighthouse at L and finds angle LAC to be 36°30'. After sailing 5 km to B, he observes angle LBC to be 73°. How many kilometers is position B from the lighthouse?

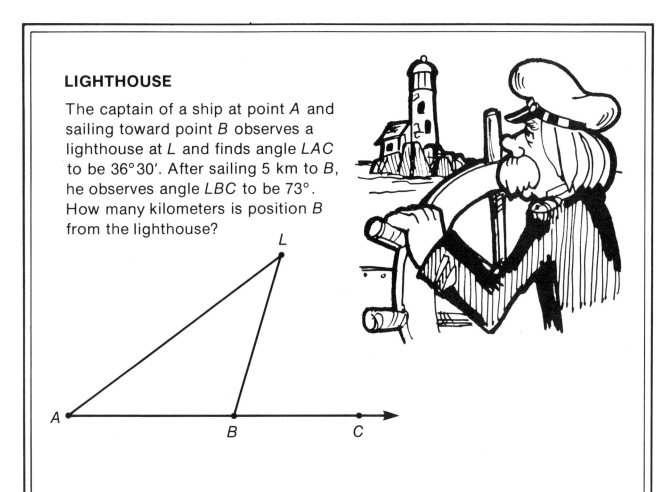

Discussion for

LIGHTHOUSE

Answer: 5 km

Clues:

1. There are 60 minutes in one degree.
2. The sum of the angle measures in a triangle is 180°.
3. The sum of the angle measures in a linear pair is 180°.
4. Classify triangle *ABL*. Is it scalene, isosceles, or equilateral?

Solution:

The two angles, ∠*LBC* and ∠*LBA*, form a linear pair, so ∠*LBA* is 107° because ∠*LBC* is 73°. Since the angle measures of a triangle sum to 180°, you can draw the following conclusions.

$$m(\angle ALB) = 180° - [m(\angle LAB) + m(\angle LBA)]$$
$$= 180° - [36°30' + 107°]$$
$$= 36°30'$$

In other words, angle *ALB* and angle *LAB* are congruent and triangle *LBA* is an isosceles triangle. The distance from position *B* to the lighthouse is $\overline{AB}$ on the diagram. Since $\overline{AB}$ is congruent to $\overline{LB}$, and the distance from *L* and *B* is 5 km, the distance from *A* and *B* is 5 km.

Teaching Suggestions:

The measure of an exterior angle of a triangle equals the sum of the measures of the two remote interior angles of the triangle. If students have learned this theorem (sometimes called the exterior angle theorem), you may wish to suggest its use. If they have not yet encountered the theorem, you may wish to use the intermediate results of this problem to lead students into deriving the exterior angle theorem.

An accurate picture helps students understand the problem. In fact, having students draw their own diagrams from the word description leads them into thinking about the relationships involved.

WARM-UP FOR PROBLEM 8

AT THE RITZ

The math club of Ritzy High designed a pennant for the school yacht. The pennant was in the shape of an isosceles triangle. Two points, P and Q, are located so that AC = AP = PQ = QB. Find the measure of angle B.

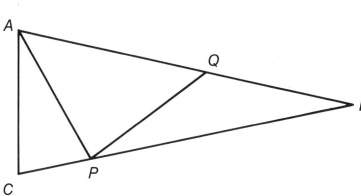

Discussion for
AT THE RITZ

Answer: 25-5/7° or about 25°43′

Clues:

1. The sum of the angles in any triangle is 180°.
2. How many of the triangles in the figure are isosceles?
3. Suppose m stands for the measure of ∠B. Try to represent the measures of each of the other angles in terms of *m*.
4. The degree measures of the angles along a straight line add to 180°.

Solution:

Let m stand for the measure of ∠B. Since triangle QPB is isosceles (QB = QP), the measure of ∠QPB is *m*, too.

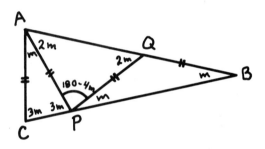

Angle PQA is exterior to triangle QPB, so ∠PQA measures *m* + *m*, or 2*m*. (The measure of an exterior angle of a triangle equals the sum of the measures of the two remote interior angles.) Triangle QPA is isosceles because QP = PA, so ∠PAQ measures 2*m*, and the third angle of triangle QPA, namely ∠QPA, has degree measure 180 − 4*m*.

CPB is a straight line segment, so the degree measures of the three angles at P together add to 180°. Therefore, ∠APC measures 3*m*.

m(∠QPB) + m(∠QPA) + m(∠APC) = 180

m + (180 − 4m) + m(∠APC) = 180

m(∠APC) = 180 − m − (180 − 4m)
     = 3m

But, triangle CAP is isosceles (PA = AC), so ∠C also measures 3*m*.

The pennant is in the shape of an isosceles triangle with AB = BC, so its base angles have equal measures and ∠BAC is congruent to ∠BCA which implies that ∠PAC must measure *m*.

The degree measures of triangle ABC sum to 180°, so the problem solution can be completed as follows.

m(∠ABC) + m(∠BCA) + m(∠BAC) = 180

m + 3m + 3m = 180

7m = 180

m = 25-5/7°

≈ 25° 43′

Teaching Suggestions:

Once students get a clear picture of the problem and see the isosceles triangles involved, they have very little trouble finding the measures of the other angles. Be ready for some spirited answering if you work the problem in class. Students will see a lot of paths to a final solution.

STEAMER STRIPE

The pennant of a steamship company is the usual isosceles triangle. The narrow end has an angle of 20° and the opposite side is 20 cm long. A blue stripe runs from one of the other corners to a point on the edge that is 20 cm from the narrow end. Determine the angle the stripe makes with the edge of the pennant.

Discussion for

STEAMER STRIPE

Answer: 150° (or its supplement)

Clues:

1. Try auxiliary lines.
2. Draw triangle ABC so that ∠A is the 20° angle and BC represents the 20 cm length. Draw $\overline{DC}$ so that D lies on $\overline{AB}$ and $\overline{AD}$ is congruent to $\overline{BC}$. (See diagram in the Solution.)
3. From D, draw a line through $\overline{AC}$ that is parallel to $\overline{BC}$.
4. Along the parallel from Clue 3, construct triangle ADE so it is congruent to triangle ABC.
5. Draw $\overline{CE}$ and find the measure of angle CED.
6. Look for congruent line segments.

Solution:

In the figure below, triangle ABC is the given triangle with ∠A measuring 20° and BC = AD = 20. $\overline{DE}$ is parallel to $\overline{BC}$, angle DEA measures 20°, and triangle ADE is congruent to triangle ABC.

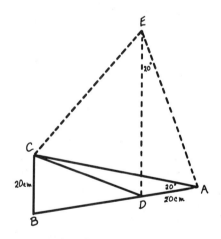

Since triangle ABC is isosceles and the degree measures of the triangle's angles sum to 180°, the measure of ∠B is 80°. Similarly, the measure of ∠EDA is 80°. By angle measure addition, the measure of ∠EAC is 60°.

By construction EA = CA, so ∠ECA is congruent to ∠CEA. Using the Angle Sum Theorem for a triangle, you can conclude that ∠ECA and ∠CEA measure 60° and, therefore, triangle CEA is equilateral.

Transitivity of congruences implies that $\overline{EC}$ and $\overline{ED}$ are congruent, so triangle ECD is isosceles. Since base angles of triangle ECD are congruent, the Angle Sum Theorem allows you to conclude that ∠EDC is 70°.

By angle addition, the angle the stripe makes with the edge of the pennant, ∠CDE, measures 70° + 80°, or 150°.

Teaching Suggestions:

The solution to this problem relies on the diagram and the diagram is not immediately obvious. Once students realize that an angle measure solution will not come from the initial diagram, they will probably try adding a variety of auxiliary lines. One reasonable figure is made by drawing a line from D to $\overline{AC}$ so that an isosceles triangle is formed. Students will find that a solution does not come from this new figure, but you may use it to lead into the figure suggested in the solution.

A lot of my students could not solve this problem. You may want to warn your students, so they don't get discouraged. (The solution requires a certain amount of serendipity.) Even so, the problem allows for some valuable discussion of triangle properties.

EXTENSION FOR PROBLEM 8

RICOCHET

A billiard player hits a ball with no English so that the ball hits all four cushions of a billiard table before it returns to its starting spot. Show how this might have happened.

Discussion for
RICOCHET

Answer:

Answers will vary. Diagrams should show a parallelogram with the angle of incidence congruent to the angle of reflection on each side of the table. (If the ball is at the exact center of the table, the shot is not possible.)

Clues:

1. A billiard table has a rectangular shape, about twice as long as it is wide (usually about 3.66 m × 1.87 m). There are no pockets. When a hit ball strikes the cushion (along the edge of the table), it rebounds at the same angle. In other words, the angle of incidence is congruent to the angle of reflection.

2. There is one position of the ball for which the problem cannot be solved.

3. Sketch a path the ball might have taken. Suppose that the closed path is a quadrilateral.

4. There is more than one way to approach this problem (even though the ball starts at the same spot each time).

5. Draw the diagonals of the rectangle.

Solution:

The rectangle shown below represents the billiard table. The ball is located at B. The rays show the path the ball will take.

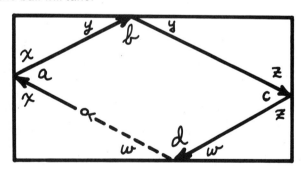

Congruent angles are marked to show that, at each cushion, the angle of incidence is congruent to the angle of reflection.

For the purposes of this solution, assume that the first ray (starting at B) and the last ray (ending at B) are collinear. Then the figure representing the path of the ball is a quadrilateral. As a result of the Angle Sum Theorem, you can write the following equations.

$x + w = 90 \qquad x + y = 90$
$z + w = 90 \qquad y + z = 90$

By solving the equations, you find that $y = w$ and $x = z$.

At each cushion, the degree measures of the three angles together add to 180°.

$2x + a = 180 \qquad 2z + c = 180$
$2y + b = 180 \qquad 2w + d = 180$

Using these equations and the information obtained from solving the previous set of equations you can conclude that $b = d$ and $a = c$. That is, in the quadrilateral representing the path of the ball, both pairs of opposite angles are congruent. Therefore, the quadrilateral is a parallelogram.

This argument suggests that, if you can construct a parallelogram that contains B (the position of the ball) so that each vertex is on a different side of the table, you can solve the problem.

A quick way to construct such a parallelogram is to draw it so that its sides are parallel to the diagonals of the table.

Another way to construct the parallelogram is by reflections. The following diagram shows the billiard table extended by a series of reflections. Successful ways to make the ball return to its starting position can be determined by drawing a straight line from the ball to one of the reflected balls.

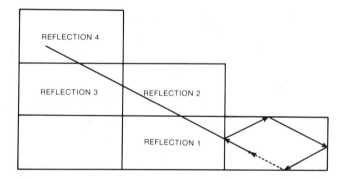

The path through each reflected table usually shows a part of the path your ball will take on the billiard table. The line drawn through Reflection 1 shows a reflection of the path the ball will take from the first cushion to the second cushion; the line through Reflection 2 shows a reflection of the path from the second cushion to the third; and so on.

By using properties of parallel lines and the Angle Sum Theorem, you can quickly show that the angles of incidence and reflection are congruent. The actual parallelogram on the table can be drawn by copying angles from the reflections.

CONTINUED ON PAGE 151

WARM-UP FOR PROBLEM 9

SHORTCUT

What is the shortest distance Nellie Bell Evans can travel?

- She is 5 km south of a stream that flows due east.
- She is 8 km west and 6 km north of her cabin.
- She wishes to water her horse at the stream and then return home.

Discussion for

SHORTCUT

Answer: $8\sqrt{5}$ or about 17.9 km

Clues:

1. The shortest distance between two points is a straight line.
2. Suppose Nellie Bell's cabin were located on the oppposite side of the stream.
3. Draw a diagram. Include a picture of the cabin on the other side of the stream.
4. How could you use reflections to solve this problem?

Solution:

Suppose that Nellie Bell's cabin were on the opposite side of the stream—a mirror reflection of its actual position. Then, because the shortest distance between two points is a straight line, the shortest path for Nellie Bell is a direct line path from her initial position to the reflected cabin's position with a stop where the line intersects the stream. The diagram below describes the situation.

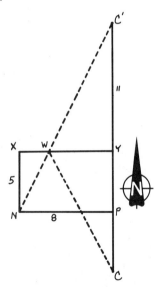

Nellie Bell's trip to the reflected cabin (at C') is the same distance as a trip to the actual cabin that waters her horse at W. The reason: a reflection is an isometry and isometries preserve distance (so that $WC' = WC$).

Note that any other path Nellie Bell might choose could be reflected through $\overline{XY}$ (the stream) and would result in a path to C' that is not a straight line and, therefore, longer than the trip first described. In other words, the shortest distance Nellie can travel is from N to W to C as shown in the diagram. (The angle of incidence to the stream is congruent to the angle of reflection from the stream; $\angle XWN \cong \angle YWC$.)

Finding the distance traveled becomes a simple application of the Pythagorean Theorem once you notice that $\overline{NC'}$ is the hypotenuse of a triangle, $\triangle NC'P$.

$$\begin{aligned}
NW + WC &= NW + WC' \\
&= NC' \\
&= \sqrt{NP^2 + C'P^2} \\
&= \sqrt{NP^2 + (C'Y + YP)^2} \\
&= \sqrt{8^2 + (11 + 5)^2} \\
&= \sqrt{320} \\
&= 8\sqrt{5} \\
&\approx 17.9
\end{aligned}$$

The shortest distance Nellie Bell can travel is approximately 17.9 km.

Teaching Suggestions:

Without reflections, the solution to this problem is quite difficult. Allow students to "reflect" on the problem for a while trying several specific positions at the watering spot and doing some guess and check before you give them any carefully directed clues. That way, they will appreciate the simplicity of the solution and will remember a new and valuable technique. However, you must not allow students to become so frustrated that they are angered by being given a "trick" problem.

The problem could have been solved equally well by reflecting Nellie Bell rather than the cabin. Some students may try that approach.

In analyzing the results of the problem, you may wish to calculate XW and WY. It's a good exercise in using similar triangles.

GEOMETRY PROBLEM 9

ONLY WITH MIRRORS

Billiards is played with only three balls—two white cue balls (distinguishable) and one red object ball—on a pool table that has no pockets. One way to score at billiards is to hit your cue ball so that it hits three cushions (sides), then hits the object ball, and then hits the opponent's cue ball. Suppose the balls for a billiard shot are in the position shown in the diagram. Show *exactly* where the cue ball must hit the first cushion so that a successful shot will be made. (Assume there is no English [spin] on the cue ball.)

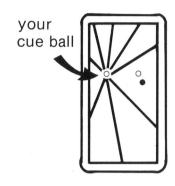

Discussion for
ONLY WITH MIRRORS

Answer:

The paths leading to Table 1 and Table 15 shown in the diagram below will give a successful shot.

Clues:

1. Draw several reflections of the billiard table on each side of the table.
2. Draw a straight line from your cue ball to the reflected balls. The point of intersection with the cushion of your table is where your cue ball should hit to make a successful one-cushion shot. Why?
3. Reflect each reflected table again to find the two-cushion shots. Reflect them again to find the three-cushion shots.
4. Not all of the straight lines you draw will give you successful shots. How can you tell which ones will?

Solution:

The diagram below shows the billiard table extended in every direction by a series of mirror reflections. The object ball and the opponent's cue ball are placed correctly in each reflection. All the possible successful shots can be determined by drawing a straight line from your cue ball (marked with a *C*) to one of the reflected balls, aiming at a point where a simple carom can be made (one ball will hit the other).

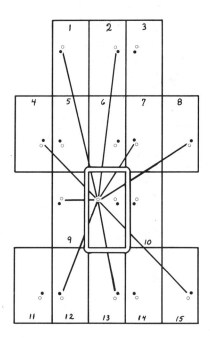

The path through each reflected table actually shows, in reflection, a part of the path your cue ball will take on the actual table. For example, the following diagram shows the path your ball will follow if you aim at Table 2. The line drawn through Table 6 shows a reflection of the path your ball will take from the first cushion to the second cushion; the line drawn through Table 2 shows a reflection of the path your ball will take from the second cushion to your opponent's cue ball.

The large diagram shows fifteen different reflected tables. Although there are other reflections, in most of them, the object balls are screened by the balls on a nearer table (the balls on Table 3 are screened by the balls on Table 6, for example), so they cannot be used to plot a successful shot.

The lines drawn show the actual direction in which the cue ball must be hit on the table. The number of cushions or reflected cushions crossed by the line of aim shows the number of cushions that your cue ball will hit before striking the balls. So, the shots into Tables 9 and Table 13 show one-cushion shots; the lines into Tables 2, 7, and 12 show two-cushion shots. Only the lines into Tables 1, 4, 8, and 15 show successful three-cushion shots.

Teaching Suggestions:

Once the reflected tables have been drawn, the problem is essentially solved. Be sure students understand why drawing lines to the reflected balls works.

I find that many students find this problem exciting. Be prepared for lots of discussion. Be sure to point out that the table in question is frictionless; the speed of the ball is thus irrelevant. Another essential part of this problem is the assumption that no English (spin) is put on the ball.

With some of my classes, I present the picture of the reflected tables along with the problem. I ask the students to discover which tables will yield solutions for one-cushion shots, which will yield solutions for two-cushion shots, and so on. For students who really enjoy this problem, I suggest they try to find the four-cushion shots.

EXTENSION FOR PROBLEM 9

TANGRAMS

Separate square *ABCD* into 7 parts using only a compass and straightedge by following these steps.

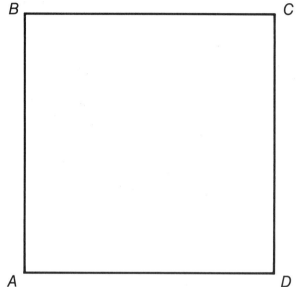

1. Draw diagonal *AC*.
2. Draw *EF* where *E* is the midpoint of *AB* and *F* is the midpoint of *BC*.
3. Draw *GD* where *G* is the midpoint of *EF*.
4. From point *E*, construct a line segment perpendicular to *AC*.
5. Construct a line segment from point *G* to *AC* that is parallel to *BC*.

Shown below are six of the thirteen convex polygons that can be made using the seven tangram pieces you have just constructed. Show how to form these shapes with the pieces and find the seven other convex polygons that can be formed.

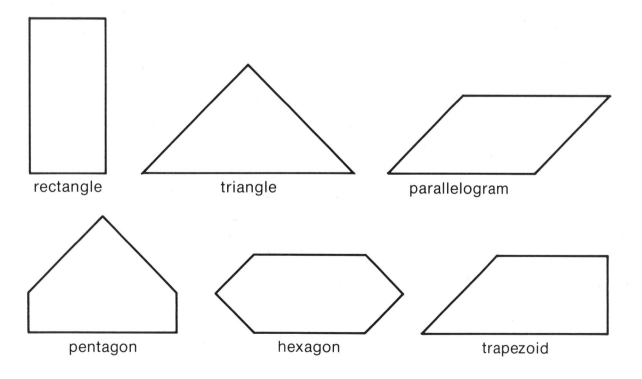

rectangle   triangle   parallelogram

pentagon   hexagon   trapezoid

WARM-UP FOR PROBLEM 10

Discussion for
TANGRAMS

Answer:

Arrangements may vary. One set of arrangements is as follows.

  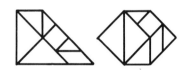

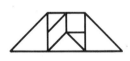 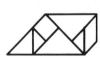

Clues:

1. Cut out the tangram shapes and physically rearrange them.
2. Try to use parts of the shapes from one figure to help create the other figures.
3. One of the seven remaining closed figures is the original square.
4. Two of the seven remaining figures are trapezoids, one of which is isosceles.
5. Of the seven remaining figures, three have six sides, and one has five sides.

Solution:

The figures can be formed in more than one way. For example, some alternate figures can be formed as follows.

 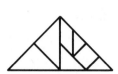

Teaching Suggestions:

The tangram is a well-known geometric puzzle that originated in China over 4000 years ago. These seven pieces can be assembled in a number of ways to form a variety of figures. Many books that give the background of tangrams and suggest different figures to make are available (for example, *Tangramath* by Dale Seymour. Palo Alto, CA: Creative Publications, 1973). You may wish to consult these books for further classroom activities or to refer to interested students for extra reading.

WARM-UP FOR PROBLEM 10

POPSICLE STICK CONSTRUCTIONS

Show how to complete each of these constructions using only an unmarked straightedge that has parallel sides.

- Construct three or more equally spaced parallel lines.
- Construct a segment twice as long as a given segment.
- Bisect a given segment.
- Bisect a given angle.
- Construct a line through a given point and parallel to a given line.
- Construct a line through a given point and perpendicular to a given line.

Discussion for

POPSICLE STICK CONSTRUCTIONS

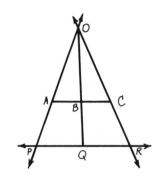

Answer: See Solution.

Clues:

1. Use a projection point when doubling a segment that is narrower than the straightedge.
2. Make sure you consider segments that are both longer than and shorter than the width of the straightedge.
3. The diagonals of a rhombus bisect one another.
4. What are some other properties of a rhombus or a parallelogram that you could use?
5. What do you know about alternate interior angles?

Solution:

PARALLEL LINES

Since the sides of the straightedge are parallel, two parallel lines can be constructed by drawing a line along each edge (E_1 and E_2). To construct a third parallel, move the straightedge so that the first edge (E_1) aligns with the line drawn along the second edge. Then draw a line along E_2. Repeat this process to construct as many equally-spaced parallel lines as required. (The distance between any pair of these lines is the width of the straightedge.)

DOUBLED SEGMENT

If the given segment $\overline{AB}$ has the same width as your straightedge, you can construct the new segment by first drawing parallel lines through A and B, then moving the straightedge to draw a third line parallel to the first two. Extend $\overline{AB}$ to a point C on the third line. Then $2AB = AC$. If the given segment is not the same width as the straightedge, first construct a line $\mathcal{L}$ that is parallel to $\overline{AB}$. Mark off three points on the line (P, Q, and R) that are equidistant ($PQ = QR =$ width of straightedge). Draw $\overleftrightarrow{PA}$ and $\overleftrightarrow{QB}$. Let O stand for the intersection. Then draw $\overline{RO}$. Next extend $\overline{AB}$ to $\overline{RO}$ at C. Since similar triangles are formed ($\triangle ABO \sim \triangle PQO$ and $\triangle BCO \sim \triangle QRO$), you can prove that $AB = BC$, which implies that $2AB = AC$.

BISECT A SEGMENT

This construction is essentially the reverse of the previous construction. Suppose $\overline{AC}$ is the given segment. Draw a line parallel to $\overline{AC}$ and mark off three points (P, Q, and R) that are the width of the straightedge apart. Draw $\overleftrightarrow{PA}$ and $\overleftrightarrow{RC}$. Let O stand for their intersection. Then draw $\overline{QO}$. $\overline{QO}$ will intersect $\overline{AC}$ at B which is the midpoint of the segment (by similar triangles).

When the length of the given segment is greater than the width of your straightedge, you can draw two pairs of parallel lines running through the endpoints. The lines will intersect, forming a rhombus. By connecting the two vertices of the rhombus that are not on $\overline{AB}$, you will bisect $\overline{AB}$. (Diagonals of a rhombus are perpendicular bisectors.)

When the length of the given segment equals the width of your straightedge, draw two lines parallel to $\overline{AC}$, one above and one below. Then draw two parallels, one through A and one through C, forming a rectangle. Finally, draw in either diagonal; they will intersect $\overline{AC}$ at its midpoint.

BISECT AN ANGLE

Draw two lines parallel to each ray of the angle. The lines will intersect at a point P, forming a rhombus. Connect that point to the vertex of the angle. This connector is a diagonal of the rhombus and therefore bisects the angle.

A LINE THROUGH A POINT AND PARALLEL TO A GIVEN LINE

Suppose P is the given point and $\mathcal{L}$ is the given line. On $\mathcal{L}$ mark off two points, A and B. Draw $\overline{PB}$. Locate the midpoint M of $\overline{PB}$ (by previously described construction methods). Draw $\overline{AM}$. Locate C on $\overline{AM}$ so that $\overline{AM} = \overline{MC}$ and M is between A and C. Draw $\overleftrightarrow{PC}$. $\overleftrightarrow{PC}$ is parallel to $\mathcal{L}$ because the two triangles formed, $\triangle PMC$ and $\triangle BMA$ are congruent (side-angle-side using construction and vertical angles) and, hence, alternate interior angles, $\angle PCM$ and $\angle BAM$, are congruent.

CONTINUED ON PAGE 153

GEOMETRY PROBLEM 10

RIGHT ON

Using only a straightedge and compass, show how to construct a right triangle whose hypotenuse has length *AB* and whose legs together have length *XY*.

A ———————————— B

X ————————————— Y

Discussion for
RIGHT ON

Answer: See Solution.

Clues:

1. How does XY compare to AB? (Is it possible that XY ≤ AB?)
2. Start with $\overline{XY}$.
3. What do you know about angles inscribed in a semicircle?
4. An isosceles triangle plays a part.
5. Construct a 45° angle at Y. (How do you do it?)

Solution:

Follow these steps.

1. Copy $\overline{XY}$.
2. At Y, construct a 45° angle, ∠XYZ.
3. From X, mark an arc with radius AB intersecting YZ at C. Draw $\overline{XC}$.
4. Find the midpoint, M, of $\overline{XC}$.
5. From M, draw an arc with radius MC intersecting $\overline{XY}$ at D. Draw $\overline{CD}$.

Triangle XCD is the required triangle.

By construction, $\overline{XC} \cong \overline{AB}$. Also △XCD is a right triangle with hypotenuse $\overline{XC}$ because an angle inscribed in a semicircle (∠CDX) is a right angle.

The two angles at D form a linear pair and, since ∠CDX is a right angle, so is ∠CDY. Since ∠CYX is a 45° angle, by the Angle Sum Theorem, ∠YCD must also be a 45° angle. Therefore ∠CYX ≅ ∠YCD, which implies that $\overline{DY} \cong \overline{CD}$ (by the converse of the Isosceles Triangle Theorem). Finally, then, XD + CD = XD + DY = XY.

Teaching Suggestions:

There are several ways to complete this construction. For example, rather than creating the semicircle in Steps 4 and 5, you could drop a perpendicular from C to $\overline{XY}$. I like to use the semicircle approach, though. This property of semicircles is extremely useful and I look for ways to point out its usefulness.

You cannot select lengths for AB and XY entirely at random. Some of them will not allow you to construct a right triangle. You might wish to work backwards from Pythagorean triples, to be sure the segments you give students will, in fact, work.

EXTENSION FOR PROBLEM 10

THE SHADOW KNOWS

The sun rises at 5:00 AM and is directly overhead at 11:00 AM. At what time in the morning to the nearest minute does a 42-m tall redwood tree cast a 50-m shadow? A person standing next to the tree casts a 1.7-m shadow at the same time that the tree is casting the 50-m shadow. Find the height of the person.

Discussion for

THE SHADOW KNOWS

Answer: The time is 7:40 AM and the person is 1.428 m tall.

Clues:

1. Corresponding sides of similar figures are proportional.
2. What is the angle of elevation to the sun?
3. The tangent of an angle is the ratio of the side opposite the angle to the side adjacent to the angle.
4. The angle of elevation of the sun changes at a constant rate.
5. If the angle of elevation to the sun were 45°, it would be 8:00 AM.
6. Draw an accurate picture illustrating the problem and use a protractor to measure the angle of elevation.

Solution:

To find the time of day, the angle of elevation to the sun, θ, must first be found. By calculating the tangent of the angle, you can find the angle.

$$\tan \theta = \frac{42}{50}$$
$$= 0.84$$
$$\theta \approx 40°$$

The sun's angle of elevation changes at a constant rate, going from 0° at 5:00 AM to 90° at 11:00 AM (90° in six hours). You need to find the amount of time it takes to change 40°. Solving the proportion

$$\frac{t}{6} = \frac{40}{90}$$

where t represents the time elapsed after 5:00 AM gives the result—t is $2.\overline{6}$ h. The time would be 7:40 AM.

The ratio of the person's height to his shadow is proportional to the ratio of the tree's height to its shadow. The problem involves solving the proportion

$$\frac{42}{50} = \frac{x}{1.7}$$

for x, where x stands for the person's height. Solving the proportion gives $x = 1.428$, so the height of the person is 1.428 m.

Teaching Suggestions:

Some students will only be able to solve the second half of this problem easily. If they have not yet had an introduction to trigonometry as ninth graders, suggest that they try a less formal approach to the problem. Have them draw an accurate model of this problem and then measure the angle of elevation using a protractor.

WARM-UP FOR PROBLEM 11

PHONE LINES

A pair of telephone poles is supported by two cables. A cable line runs from the top of each pole to the bottom of the other. If one pole is 20 m high and the other pole is 30 m high, how high above the ground do the cables intersect?

Discussion for

PHONE LINES

Answer: 12 m

Clues:

1. Draw a diagram of the situation and look for similar triangles.
2. What proportions can you set up using similar triangles?
3. If $a/b = c/d$ and $w/x = y/z$, then $(a/b) + (w/x) = ??$
4. In the figure, $\overline{EF}$ separates the segment between the two poles into two parts. What ratio do the measures of these parts form?

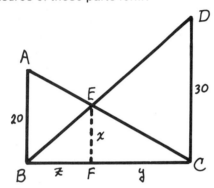

5. You don't need to know the distance between the two poles.

Solution:

There are several ways to solve this problem. Here are two.

SOLUTION 1: In the figure shown in Clue 4, $\overline{AB}$ and $\overline{DC}$ represent the telephone poles; $\overline{BC}$ is the distance between them. The two cables, $\overline{AC}$ and $\overline{DB}$, intersect at E; x represents the height above the ground. $\overline{EF}$ separates $\overline{BC}$ into two parts; the ratio of these parts is proportional to the ratio of $\overline{AB}$ to $\overline{DC}$. In other words, since $AB:DC = 20:30$, or $2:3$, the ratio of z to y is $2:3$ and the ratio $BC:z$ is $5:2$.

As a result, you can write and solve a proportion involving x using $\triangle DCB \sim \triangle EFB$.

$$\frac{BC}{z} = \frac{DC}{x}$$

$$\frac{5}{2} = \frac{30}{x}$$

$$x = 12$$

The 2:3 ratio comes from observing that triangle ABC is similar to triangle EFC and that triangle DCB is similar to triangle EFB. The similarity of triangles EFC and ABC implies that

$$\frac{x}{y} = \frac{20}{y+z}$$

The similarity of triangles EFB and DCB implies that

$$\frac{x}{z} = \frac{30}{y+z}$$

Solving both of these proportions gives

$$y + z = \frac{20y}{x} \quad \text{and} \quad y + z = \frac{30z}{x}$$

$$\frac{20y}{x} = \frac{30z}{x}$$

$$20y = 30z$$

The ratio $z:y$ is, therefore, 20:30, or 2:3.

SOLUTION 2: Triangles BFE and BCD are similar, so $BF:FE = BC:CD$; triangles CFE and CBA are similar, so $FC:FE = BC:AB$. Adding these two proportions (using the addition property of equality) gives

$$\frac{BF}{FE} + \frac{FC}{FE} = \frac{BC}{CD} + \frac{BC}{AB}$$

$$\frac{BF + FC}{FE} = \frac{BC}{CD} + \frac{BC}{AB}$$

But $BF + FC = BC$.

$$\frac{BC}{FE} = \frac{BC}{CD} + \frac{BC}{AB}$$

Dividing each term by BC leaves the following equation.

$$\frac{1}{FE} = \frac{1}{CD} + \frac{1}{AB}$$

This equation can be solved for FE, since the values for CD and AB are known.

$$\frac{1}{FE} = \frac{1}{30} + \frac{1}{20}$$

$$= \frac{5}{60}$$

$$FE = 12$$

Teaching Suggestions:

Emphasize that the distance between two poles is an *arbitrary* piece of information. Some students will find it helpful to assign a value, say 100, to the distance and work with the resulting proportions. Be sure they are convinced that other values for the distance will work equally well.

Solving the problem the second way allows you to go over basic properties of proportions in a meaningful way. I have found students to be very responsive to this approach.

GEOMETRY PROBLEM 11

SIDE SHOW

If the measures of the sides of a triangle are consecutive even integers and the measure of the greatest angle is twice that of the smallest angle, find the measures of the sides of the triangle.

Discussion for
SIDE SHOW

Answer: 8 units, 10 units, and 12 units

Clues:

1. Consecutive even integers can be represented by $2x$, $2x + 2$, $2x + 4$, and so on.
2. Which side of the triangle is opposite the smallest angle? The largest angle?
3. You will need an auxiliary line to solve the problem. (Bisect the largest angle.)
4. You will need to know an important property about an angle bisector of a triangle.
5. Look for proportional parts (and similar triangles).
6. Use two equations with two unknowns.
7. If one part of a segment measuring $(2x + 4)$ units measure y units, what is the measure of the other part?

Solution:

In triangle ABC, let $\angle C$ be the smallest angle and $\angle A$ the largest. The lengths of the sides are given by $AB = 2x$, $AC = 2x + 2$, and $CB = 2x + 4$. Find D on $\overline{CB}$ so that $\overline{AD}$ bisects $\angle A$. Let $BD = y$.

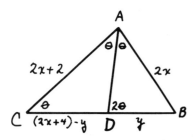

The angle bisector separates $\overline{CB}$ into segments whose lengths are proportional to $\overline{AB}$ and $\overline{AC}$.

$$\frac{AB}{BD} = \frac{AC}{DC}$$

Because $\overline{AD}$ bisects angle A, $\angle DAB \cong \angle C$. Also, since the measure of an exterior angle of a triangle equals the sum of the measures of the remote interior angles, $\angle ADB \cong \angle CAB$. These two congruences imply that $\triangle ABD \sim \triangle CBA$, yielding the following proportion.

$$\frac{AB}{BD} = \frac{CB}{AB}$$

The two proportions give two equations in two unknowns.

$$\frac{AB}{BD} = \frac{AC}{DC} \quad \text{gives} \quad \frac{2x}{y} = \frac{2x + 2}{(2x + 4) - y}$$

$$\frac{AB}{BD} = \frac{CB}{AB} \quad \text{gives} \quad \frac{2x}{y} = \frac{2x + 4}{2x}$$

Solving the second equation for y gives $y = (x^2)/(x + 4)$. Substituting this value into the first equation and simplifying yields a quadratic, $4x^2 - 16x = 0$. Only one root, 4, is a reasonable answer and, therefore, the sides are given by

$AB = 2x = 8$,

$AC = 2x + 2 = 10$, and

$CB = 2x + 4 = 12$.

Teaching Suggestions:

Most of my students start this problem by assuming the triangle must be a right triangle. Their solution under this assumption is 6 units, 8 units, and 10 units. However, they will accept my explanation for why the triangle cannot be a right triangle. The right angle would have to be the largest angle and, since it is twice the size of the smallest angle, one of the acute angles must be a 45° angle, making the triangle isosceles. But the triangle, as given, is scalene.

The key to solving this problem is in realizing that the angle bisector will provide the desired proportions. Beware. Weaker students will get bogged down in the algebra of the problem. They will need a thorough explanation on simplifying the equations.

This problem could be simplified algebraically by letting consecutive even integers be represented by x, $x + 2$, and $x + 4$ where x is even.

EXTENSION FOR PROBLEM 11

ROUND ABOUT

The sides of a right triangle are in arithmetic progression. The radius of a circle circumscribed about the triangle is 25 cm. Find the lengths of the three sides of the triangle.

Discussion for
ROUND ABOUT

Answer: 30 cm, 40 cm, and 50 cm

Clues:

1. The center of the circumscribed circle lies on the hypotenuse of the right triangle.
2. The diameter of the circumscribed circle is the same as the hypotenuse of the right triangle.
3. Numbers in arithmetic progression can be represented by x, $x + d$, $x + 2d$, and so on, where d stands for the common arithmetic difference.
4. Find values for the sides of the right triangle in terms of d. (They are x, $x + d$, and $x + 2d$.)

Solution:

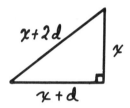

Let x and $x + d$ represent measures of the two legs of the right triangle and let $x + 2d$ represent the hypotenuse. The Pythagorean Theorem gives the following equation.

$$x^2 + (x + d)^2 = (x + 2d)^2$$

Simplify the equation and solve for x.

$$x^2 + x^2 + 2dx + d^2 = x^2 + 4dx + 4d^2$$
$$x^2 - 2dx - 3d^2 = 0$$
$$(x + d)(x - 3d) = 0$$
$$x = -d \text{ or } 3d$$

Since measures of triangle sides are not negative, $-d$ is an extraneous root. Substituting $3d$ in place of x in the original values results in sides measuring $3d$, $4d$, and $5d$, so the triangle is a 3-4-5 right triangle.

The center of the circle lies on the hypotenuse of the triangle since the radius of the circumscribed circle is 25 cm. The hypotenuse measures 50 cm. In other words, $5d = 50$ and the common arithmetic difference is 10. This value could be substituted for the original values to give sides of x, $x + 10$, and $x + 20$, but we know more; the sides are $3d$, $4d$, and $5d$, giving values of 30 cm, 40 cm, and 50 cm.

Teaching Suggestions:

Most of my students come up with the solution to this problem by assuming the triangle is a 3-4-5 right triangle. We use this triangle so often in our teaching that students come to believe that all right triangles are 3-4-5 right triangles. In this particular case, the triangle is 3-4-5, but unless my students can *show* me why, I don't consider that they have really worked the problem.

To reinforce students' use of square root tables, I sometimes ask them to find other right triangles that have a hypotenuse measuring 50 units and having legs that are integer units with the aid of tables. There is only one other that I have found; it has sides of 14 units and 48 units.

WARM-UP FOR PROBLEM 12

I NEVER PROMISED YOU . . .

Tightwad has decided to change his rectangular rosebed into the shape of a right triangle.

- He wants the new bed to have the same area as the old.
- He wants the smallest possible perimeter (so he won't need a lot of fencing).
- His rectangular rosebed now measures 24 m × 35 m.

What dimensions should Tightwad use for his triangular rosebed? (Consider integer values only!)

Discussion for
I NEVER PROMISED YOU . . .

Answer: 40 m, 42 m, and 58 m

Clues:

1. The product of the measures of the legs of the right triangle is 1680.
2. There are only 20 possible ways to obtain 1680 as the product of two numbers.
3. Remember Pythagoras.
4. The square root of 1680 is approximately 41.
5. Work from the middle out.

Solution:

The area of the rectangular rosebed is 840 m². If a and b are the measures of the legs of the triangular rosebed, then $(½) ab = 840$, so $ab = 1680$ and the only possible measurements are factors of 1680. The twenty integer combinations are:

1 · 1680	2 · 840	3 · 560	4 · 420
5 · 336	6 · 280	7 · 210	8 · 210
10 · 168	12 · 140	14 · 120	15 · 112
16 · 105	20 · 84	21 · 80	24 · 70
28 · 60	30 · 56	35 · 48	40 · 42

Of the twenty possible two-number factorizations, only three can possibly be legs of a right triangle, namely 15 · 112, 24 · 70, and 40 · 42. The sum of the squares of the terms for the other combinations are not perfect squares.

The lengths of the sides of the triangular rosebed could be:

    42 m, 40 m, and 58 m,

    70 m, 24 m, and 74 m, or

    112 m, 15 m, and 113 m.

Since the sum of the measures of the sides is the perimeter of the rosebed, the smallest perimeter is 140 m from dimensions of 42 m, 40 m, and 58 m. (The other perimeters are 168 m and 240 m.)

Teaching Suggestions:

I encourage my students to keep a calculator nearby when they solve this problem. Some students prefer to write computer programs for the problem, in which case I insist they submit copies of their programs along with their solutions.

Make sure the students realize that one side of the triangular rosebed is the hypotenuse of a right triangle. Also, be sure they understand that only integer values meet the requirements of the problem.

GEOMETRY PROBLEM 12

EQUAL RIGHTS

A garden in the shape of a right triangle has sides measuring 60 units, 80 units, and 100 units. A fence runs from the right angle to the hypotenuse and separates the garden into two parts of equal perimeter. Find the length of the fence.

Discussion for
EQUAL RIGHTS

Answer: $24\sqrt{5}$ units, or about 53.7 units

Clues:

1. The fence is not a median, an angle bisector, or an altitude.
2. Find the lengths of the two parts of the hypotenuse.
3. Draw the altitude from the fence point on the hypotenuse to the 80 unit side.
4. Look for similar triangles.
5. Use the Pythagorean Theorem to find the lengths of the two segments on the 80 unit side.

Solution:

The figure below represents the garden. The fence runs from A to D.

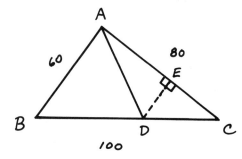

According to the problem, △ABD has the same perimeter as △ACD.

$$\text{Perimeter}(\triangle ABD) = \text{Perimeter}(\triangle ACD)$$
$$60 + AD + (100 - DC) = 80 + AD + DC$$
$$160 - DC = 80 + DC$$
$$2DC = 80$$
$$DC = 40$$

The figure shows an altitude for triangle ADC. By adding this auxiliary line to the figure, you create a triangle that is similar to triangle ABC, namely triangle EDC. (They are both right triangles and share angle C.)

By using proportions, you can find a value for EC and, hence AE. Then, by the Pythagorean Theorem, you can find a value for AD.

$$\frac{EC}{AC} = \frac{DC}{BC}$$
$$\frac{EC}{80} = \frac{40}{100}$$
$$EC = 32$$
$$AD^2 = AE^2 + ED^2$$
$$= (AC - EC)^2 + ED^2$$
$$= 48^2 + 24^2$$
$$= 2880$$
$$AD = 24\sqrt{5}$$
$$\approx 53.7$$

Teaching Suggestions:

Some of your students may have had an introduction to trigonometry as part of their first year algebra course. They may be tempted to try to use their trigonometry, but, unless they have been introduced to the Law of Cosines, they will not get very far.

I find that having my students go from the desired answer to the known information is helpful. For example, to find AD, they need to know AE and ED. To find AE and ED, they need to know EC and DC, and so on. They should keep backing up until they can reach the given information.

CROW'S FEET

A cathedral tower 200 ft high is 250 ft from a church that is 150 ft high. At the same instant, two crows, one from each tower, fly off at the same speed heading toward some grain that is on a level, straight road located between the towers. The crows reach the grain at the same instant. How far is the grain from the foot of the cathedral tower?

WARM-UP FOR PROBLEM 13

Discussion for

CROW'S FEET

Answer: 90 ft

Clues:

1. There are two right triangles involved.
2. Each crow flies the same distance.
3. If x represents the distance of the grain from the cathedral tower, how can you represent the distance of the grain from the church tower?
4. Write two different equations using two different unknowns.

Solution:

Since the crows leave at the same instant, fly at the same speed, and arrive at the same instant, they each fly the same distance.

Let y stand for the distance each crow flies and let x represent the distance of the grain from the foot of the cathedral tower. The following diagram describes the situation.

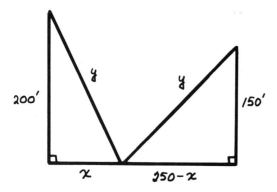

Since the triangles are right triangles, you can use the Pythagorean Theorem to make two equations.

$$y^2 = 200^2 + x^2 \text{ and } y^2 = 150^2 + (250 - x)^2$$

Use substitution to solve the equations for x.

$$40{,}000 + x^2 = 22{,}500 + 62{,}500 - 500x + x^2$$
$$500x = 45{,}000$$
$$x = 90$$

Teaching Suggestions:

Once students recognize that there are two right triangles having the same hypotenuse, they can easily set up the equations. Drawing a correct diagram and representing the distance between the two buildings by x and $(250 - x)$ are keys to this problem.

WARM-UP FOR PROBLEM 13

LOONEY TUNES

One circle is constructed on each side of a right triangle.

- The center of each circle is the midpoint of the side and the side forms a diameter of the circle.
- The area of the triangle is 24 square units.

Find the total area of the regions of the two smaller circles that lie *outside* the largest circle.

Discussion for
LOONEY TUNES

Answer: 24 square units

Clues:

1. The formula for the area of a circle is $A = \pi r^2$.
2. The largest circle contains each of the three vertices of the triangle.
3. Express the area of the regions in terms of semicircles and triangles.

Solution:

You can write two different equations about Semicircle$_3$ shown in the figure below.

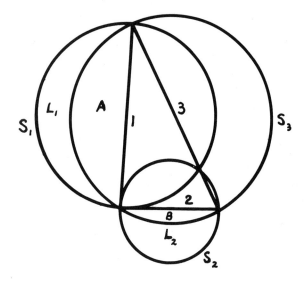

Area (Semicircle$_3$) = Area (A) + Area (B) + Area (Triangle)

Area (Semicircle$_3$) = Area (Semicircle$_1$) + Area (Semicircle$_2$)
$\qquad$ = Area (L_1) + Area (A) + Area (L_2) + Area (B)

Then, you can equate the two expressions.

Area (L_1) + Area (A) + Area (L_2) + Area (B) =
$\qquad$ Area (A) + Area (B) + Area (Triangle)

Area (L_1) + Area (L_2) = Area (Triangle)

Another, less elegant way to solve the problem is as follows.

To find the area of the two moon-shaped regions, first find the areas of the two semicircles on the legs of the right triangle. Then subtract the areas of the unshaded parts of the semicircles; they equal the area of half the large circle minus the area of the triangle.

Let a and b stand for the measures of the legs of the right triangle and h stand for the measure of the hypotenuse. Then $h^2 = a^2 + b^2$ and the problem is solved as follows.

$$\begin{aligned}\text{Area of Shaded Regions} &= \left[\text{Area of Semicircle}_a + \text{Area of Semicircle}_b\right] - \left[\text{Area of Semicircle}_h - \text{Area of Triangle}\right] \\ &= \left[\frac{1}{2}\left(\frac{\pi a^2}{4}\right) + \frac{1}{2}\left(\frac{\pi b^2}{4}\right)\right] - \left[\frac{1}{2}\left(\frac{\pi h^2}{4}\right) - \frac{1}{2}(ab)\right] \\ &= \frac{\pi a^2 + \pi b^2 - \pi h^2 + 4ab}{8} \\ &= \frac{\pi(a^2 + b^2) - \pi(a^2 + b^2) + 4ab}{8} \\ &= \frac{4ab}{8} \\ &= \frac{1}{2}ab \\ &= \text{Area of Triangle} \\ &= 24\end{aligned}$$

Teaching Suggestions:

A large part of the work in solving this problem is in isolating the various parts of the figure and seeing which parts together help you find the area. Drawing several sketches of the figure and coloring parts sometimes helps students visualize the relationships.

Probably no other problem in mathematics has had a greater or longer attraction than that of squaring the circle (constructing a square equal in area to a given circle). Since as far back as 1800 BC, people have worked on the problem. Hippocrates (around 440 BC) succeeded in squaring certain special lunes, moon-shaped figures bounded by two circular arcs. Did he hope that his work might lead to a solution of the square problem?

Looney Tunes is one of Hippocrates' investigations. In fact, the theorem reporting the results of the *Looney Tunes* problem is sometimes called the Lunes of Hippocrates theorem.

In more recent times, it has been proved that squaring the circle cannot be accomplished with a straightedge and compass alone, but still somehow there is a fascination with the problem and every year someone tries a new solution.

GEOMETRY PROBLEM 13

ALL WET

A rectangular lawn has an area of 126 m². Surrounding the lawn is a flower border 4 m wide having an area of 264 m². A circular sprinkler is installed in the middle of the lawn. What is the spraying radius of the sprinkler if it covers the entire yard (including the flower border)?

Discussion for

ALL WET

Answer: Approximately 15 m $\left(\frac{1}{2}\sqrt{901} \text{ to be exact}\right)$

Clues:

1. Find the dimensions of the lawn.
2. Let x and y represent the dimensions of the lawn. What are the dimensions of the entire yard?
3. There are no fractions involved—in dimensions of the yard, that is.
4. The sum of the width and length of the lawn is 25.
5. Find two equations in two unknowns, x and y. Use substitution to solve them.
6. You will need to solve a quadratic equation.

Solution:

Let x and y represent the length and width of the lawn, respectively. Then $x + 8$ and $y + 8$ represent the dimensions of the entire yard.

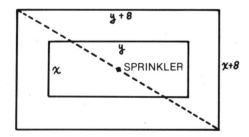

The area of the lawn is 126 m². So $xy = 126$ and, since the area of the flower border is 264 m², $(x + 8)(y + 8) - xy = 264$. The second equation simplifies as follows.

$$xy + 8y + 8x + 64 - xy = 264$$
$$8(y + x) = 200$$
$$y + x = 25$$
$$y = 25 - x$$

Substituting $25 - x$ for y into the first equation gives the following results.

$$xy = 126$$
$$x(25 - x) = 126$$
$$25x - x^2 = 126$$
$$x^2 - 25x + 126 = 0$$
$$(x - 18)(x - 7) = 0$$
$$x = 18 \text{ or } 7$$
$$y = 7 \text{ or } 18$$

EXTENSION FOR PROBLEM 13

The lawn is 18 m × 7 m. The entire yard is 26 m × 15 m.

Since the sprinkler is centered in the lawn, the greatest radius it must cover is along the diagonal of the yard. The Pythagorean Theorem gives the length of the diagonal.

$$(\text{diagonal})^2 = 26^2 + 15^2$$
$$= 901$$
$$\text{diagonal} = \sqrt{901}$$
$$\approx 30.02$$

The sprinkler must be located at the midpoint of the diagonal, making its spraying radius about 15 m from each corner of the yard.

Teaching Suggestions:

Because this problem is basically algebraic in nature, I find that my students need the review of first-year algebra. In most cases, students are able to solve this problem and find it rewarding when they discover that they can solve some of the more difficult types of word problems that gave them troubles the year before.

Emphasize that, since two areas are given, two equations involving these values are needed. Once students have found these two equations, they usually realize that solving the equations by the substitution method is the way to proceed.

DIAMOND IN THE ROUND

In the diagram below, HCEG is a rhombus formed by connecting the midpoints of a rectangle ABDF. Find the length of a side of the rhombus if OG = 10 and CK = 8. Assume that O is the center of the circle circumscribing rectangle ABDF.

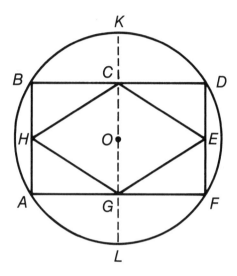

Discussion for
DIAMOND IN THE ROUND

Answer: 18

Clues:

1. What is the radius of the circle?
2. Separate the rectangle into four congruent parts.
3. The diagonals of a rectangle are _____.
4. Draw $\overline{OB}$.

Solution:

Drawing $\overline{HE}$ separates the rectangle into four smaller congruent rectangles. Each side of the rhombus is a diagonal of one of the rectangles. But, the diagonals of a rectangle are congruent so, for example, $\overline{HC}$ is congruent to $\overline{OB}$, a radius of the circle.

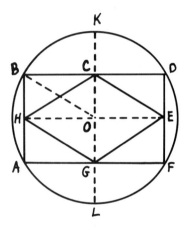

The segment $\overline{OK}$ is a radius radius of the circle and $OK = OC + CK$. Also, $OG = OC$. Therefore, $OK = OG + CK = 10 + 8 = 18$. The sides of the rhombus are 18 units long.

Teaching Suggestions:

This problem becomes trivially easy once students discover that the sides of the rhombus are congruent to radii of the circle. Normally, I present problems like this one at the end of a class period, so students quick to see the solution don't ruin it for the others.

I like to have my students find the dimensions of the given rectangle, too. The Pythagorean Theorem works nicely here, but I like students to try the exercise without it. I have them think in terms of intersecting chords.

WARM-UP FOR PROBLEM 14

A RIVER TOO WIDE

How wide is the river?

- A bridge across the river is built in the shape of a circular arc.
- The middle of the bridge is ten meters above the water.
- Twenty-seven meters from shore, the bridge is nine meters above the water.

Discussion for
A RIVER TOO WIDE

Answer: 80 m

Clues:

1. Do *not* try similar triangles.
2. You will need the Pythagorean Theorem.
3. Try to define two variables and write a system of quadratic equations using them.
4. Look at these two diagrams. Find FE, FC, and CE. Also find AG, GC, and AC.

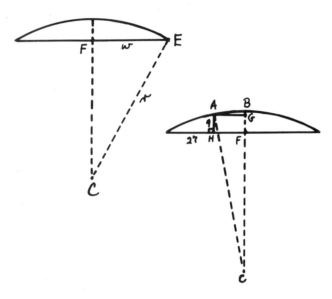

Solution:

Let r stand for the radius of the circle from which the bridge is built. Let $2w$ stand for the width of the river. (Then w is the distance from the shore to the middle of the river.)

The diagram below shows a cross section of the river and bridge; $\overline{DE}$ represents the river and $\widehat{DE}$ the bridge.

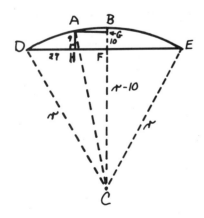

Two triangles, $\triangle CFE$ and $\triangle ABC$ give information that can be used to solve the problem. Applying the Pythagorean Theorem to $\triangle CFE$ gives $w^2 + (r - 10)^2 = r^2$. Applying it to $\triangle AGC$ gives $(w - 27)^2 + (r - 1)^2 = r^2$.

The first equation simplifies to:

$$w^2 = 20r - 100$$

The second equation simplifies to:

$$2r = w^2 - 54w + 730$$

Substituting $10(w^2 - 54w + 730)$ for $20r$ in the first equation gives the following results.

$$w^2 = 10(w^2 - 54w + 730) - 100$$
$$9w^2 - 540w + 7200 = 0$$
$$w^2 - 60w + 800 = 0$$
$$(w - 20)(w - 40) = 0$$
$$w = 20 \text{ or } 40$$

Since the distance from shore to the middle of the river is greater than 27 m, the only reasonable value for w is 40 and the river must be 80 m wide.

Teaching Suggestions:

When my geometry students first approach this problem, they try to use similar triangles. Several quick, but accurate, sketches point out the problems with this approach.

I usually have to walk students through this problem day by day. The diagrams in Clue 4 are really a final big hint. The second diagram, in particular, gets students going.

Some people get stuck by wanting to use $\overline{CH}$ rather than $\overline{CA}$.

You may find, as I have, that some geometry students' algebraic skills have deteriorated, and you will need to give a short refresher lesson on "FOIL" or other methods for squaring binomials. Interestingly enough, I have used this problem both in algebra classes and geometry classes; I experienced very few difficulties with it in the algebra classes.

I like the opportunity to solve quadratic systems of equations. It's rather new and an eye-opener for many students.

GEOMETRY PROBLEM 14

RADIALS PLEASE

In measuring a triangle inscribed in a circle, a student finds that one angle measures 60° and the side opposite measures 10 cm. Can you find the radius of the circumscribed circle from this information? Show how.

Discussion for

RADIALS PLEASE

Answer: $\frac{10}{3}\sqrt{3}$ cm (or approximately 5.77 cm)

Clues:

1. Is the center of the circle inside the triangle?
2. What is the measure of the arc intercepted by the 60° angle?
3. What is the measure of the central angle that intercepts the arc cut off by the 10 cm chord?
4. What type of triangle is formed when the center of the circle is used as one vertex and a chord is used as one side?
5. The measures of the acute angles in the isosceles triangle formed are 30°.
6. If all else fails, use the Pythagorean Theorem.

Solution:

Let $\triangle ABC$ represent the given triangle and let P stand for the center of the circle. Then $\overline{PA}$ and $\overline{PB}$ are radii of the circle and, thus, $\triangle APB$ is an isosceles triangle.

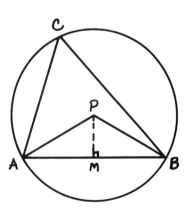

Since $\overset{\frown}{AB}$ is intercepted by $\angle ACB$, which measures 60°, the measure of $\overset{\frown}{AB}$ is 120°. As a result, the central angle, $\angle APB$, measures 120°.

Draw $\overline{PM}$ perpendicular to $\overline{AB}$. Since $\triangle APB$ is isosceles, $\overline{PM}$ bisects both $\angle APB$ and $\overline{AB}$. So, $\angle APM$ measures 60° and AM is 5. The sum of the degree measures of a triangle is 180°, so $\angle PAM$ measures 30°. In other words, $\triangle APM$ is a 30°-60° right triangle. Therefore,

$$AM = \frac{\sqrt{3}}{2} AP$$

$$5 = \frac{\sqrt{3}}{2} AP$$

$$AP = 5\left(\frac{2}{\sqrt{3}}\right)$$

$$= \frac{10}{3}\sqrt{3}$$

$$\approx 5.77$$

Teaching Suggestions:

When I use this problem with my students, I sometimes present the solution as a worksheet and ask students to give reasons for each step in the solution.

Realizing that the central angle intercepting the 10 cm side is 120° leads to drawing in the isosceles triangle and indicates a method of attack.

EXTENSION FOR PROBLEM 14

SERGEANT'S WALK

An army sergeant left her barracks walking along azimuth 330°. Upon reaching a small hill, she turned and walked along azimuth 30° until she came to a tree. Here she made a 60° right turn and continued walking until she reached a bridge. She turned to azimuth 150° and walked beside the river.

Half an hour later, she came upon a mill. Changing her direction again, she walked along azimuth 210°, her goal being a miller's house. At the house she made a right turn and walked along azimuth 270° to finish her tour.

Using a protractor, draft the sergeant's route neatly and find her ending spot. Assume she walked 2.5 km in each direction.

Discussion for
SERGEANT'S WALK

Answer: The sergeant ended where she began, at her barracks. See diagram in solution below for her route.

Clues:

1. Azimuth represents direction. Azimuths are used in flying airplanes.
2. Azimuth directions are usually measured clockwise in degrees along the horizon from a point due north (south).
3. An azimuth direction of 315° means a direction that is 45° to the west (east) of a point due north (south).

Solution:

The diagram shows the sergeant's route.

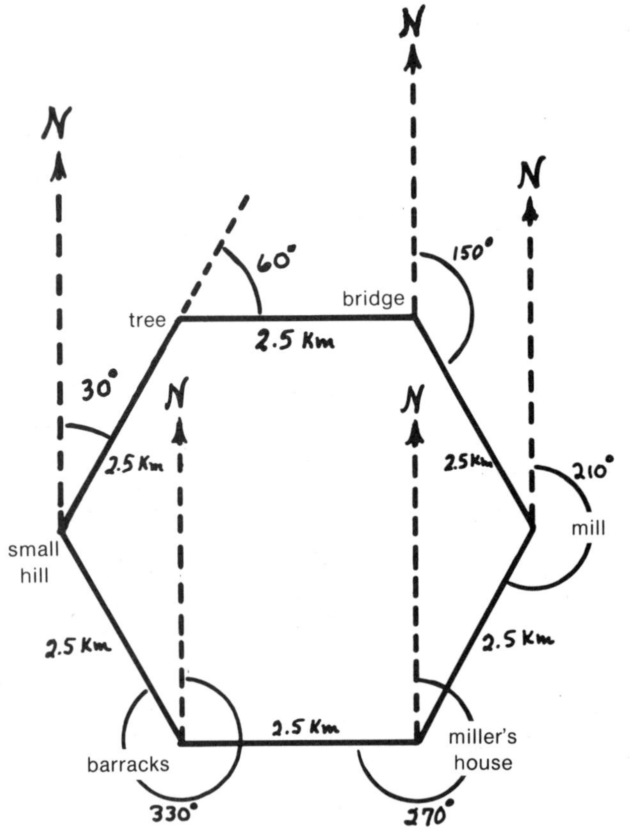

Note that wherever an azimuth measurement is given, a ray is drawn pointing directly north (south) and the angle is measured in a clockwise direction from the northern (southern) ray.

A few quick calculations based on the information given and the diagram will show that the rays formed by the sergeant's paths meet at 120° angles. Each part of the sergeant's walk was 2.5 km. Therefore, the figure shown in the diagram is, indeed, a closed figure; in fact, it is a regular hexagon. So, the sergeant ended her walk where she began—at the barracks.

Teaching Suggestions:

Once students get the idea of azimuth, they seem to do all right on their own. The third direction, where the sergeant makes a 60° right turn, will confuse some students because it is *not* an azimuth reading.

You will notice that all the angle measurements are multiples of 30°. Since this is the case, I ask my more experienced students to reconstruct the sergeant's route without a protractor—using only a compass and straightedge.

WARM-UP FOR PROBLEM 15

WALK STRAIGHT

Suppose a person walks one kilometer east, then one kilometer northeast, then another kilometer east. Find the distance, in kilometers, between the person's initial and final positions.

Discussion for

WALK STRAIGHT

Answer: $\sqrt{5 + 2\sqrt{2}}$ km (about 2.80 km)

Clues:

1. Draw a diagram using directed lines (vectors) to illustrate the problem.
2. Extend the first vector and, from the endpoint of the last vector, drop a perpendicular to this extension.
3. Use the Pythagorean Theorem.
4. If the hypotenuse of a 45°-45°-90° triangle is 1 unit long, then how long is each leg of the triangle?

Solution:

In the diagram below, S represents the person's initial starting position and F represents the final position. $\overline{SA}$ represents the first walk east; $\overline{AB}$ represents the walk northeast; $\overline{BF}$ represents the last portion of the walk.

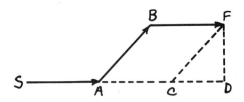

One approach to solving this problem is to extend SA to a point D where it intersects a perpendicular from F. The distance from the initial position to the final position is SF, the length of the hypotenuse of triangle FSD. To find this length, you find SD and FD, then use the Pythagorean Theorem.

Draw $\overline{FC}$ so that $\overline{FC} \parallel \overline{AB}$. Since the walker headed east from S and from B, the segments $\overline{BF}$ and $\overline{SA}$, and therefore $\overline{AC}$, are parallel. This information implies that figure $ABFC$ is a parallelogram. In fact, the figure is a rhombus because, by the problem statement, $AB = BF$. Therefore, all four sides are congruent and $AC = CF = 1$.

By construction, $\overline{AB} \parallel \overline{CF}$. Consequently, $\angle BAC \cong \angle FCD$. Because the walker headed northeast from A, the angle at B, angle BAC, measures 45°. Therefore, angle FCD measures 45° and triangle FCD is a 45°-45°-90° triangle whose hypotenuse ($\overline{CF}$) is 1 unit long. As a result, FD and CD are $\sqrt{2}/2$ units long.

$\overline{SD}$ has three parts: $\overline{SA}$, $\overline{AC}$, and $\overline{CD}$. Thus, SD is $1 + 1 + (\sqrt{2}/2)$, or $(4 + \sqrt{2})/2$.

Now, apply the Pythagorean Theorem to find SF.

$$SF^2 = SD^2 + FD^2$$
$$= \left(\frac{4 + \sqrt{2}}{2}\right)^2 + \left(\frac{\sqrt{2}}{2}\right)^2$$
$$= \frac{10 + 4\sqrt{2}}{2}$$
$$= 5 + 2\sqrt{2}$$
$$SF = \sqrt{5 + 2\sqrt{2}}$$

Teaching Suggestions:

Walking northeast means you walk at an angle of 45° from north (or 45° from east). Once students understand this simple fact, they can usually see a solution path in their diagrams.

There are a number of ways to approach this problem. Students will benefit from seeing how different people have solved the problem. If you want to present an elegant solution, use the Law of Cosines.

$$SF^2 = SC^2 + CF^2 - 2(SC)(CF)\cos(\angle SCF)$$
$$= 2^2 + 1^2 - 2(2)(1)\cos 135°$$
$$= 4 + 1 + 4\left(\frac{\sqrt{2}}{2}\right)$$
$$= 5 + 2\sqrt{2}$$
$$SF = \sqrt{5 + 2\sqrt{2}}$$

GEOMETRY PROBLEM 15

TREASURE HUNT

A small sea island has only three conspicuous landmarks—three palm trees, each 100 m from the other two. Two of the trees are along a north-south line. An old treasure map gives the following directions:

Proceed from the southernmost tree 15 m due north, then 26 m due west.

Is the treasure buried within the triangle formed by the trees?

Discussion for

TREASURE HUNT

Answer: No, the treasure is buried outside the triangle.

Clues:

1. What kind of triangle is formed by the three palm trees?
2. There are two different cases to consider.
3. What are the measures of the angles in the right triangle formed by the path to the treasure?

Solution:

There are two different possible orientations of the three trees.

CASE I: The third palm tree is east of both trees on the north-south line.

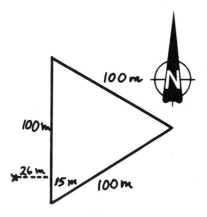

A diagram shows that the treasure lies to the west of the north-south line, so it cannot possibly be buried within the triangle formed by the trees.

CASE II: The third palm tree is west of both trees on the north-south line.

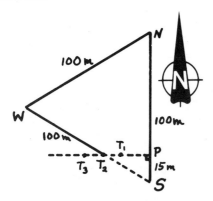

The diagram requires closer study. First, notice that the triangle formed by the three palm trees is equilateral and, hence, its three angles each measure 60°.

The treasure lies somewhere on $\overrightarrow{PT}$—either at a point inside the triangle, T_1, at a point on the triangle, T_2, or at a point outside the triangle, T_3. Since PST_2 is a right triangle with $m(\angle S) = 60°$, triangle PST is a 30°-60° right triangle. The side opposite the 60° angle has measure $\sqrt{3}$ times that of the side opposite the 30° angle. Thus,

$$PT_2 = \sqrt{3}\,(15)$$
$$\approx 25.98$$

By the properties of order, $PT_1 < PT_2 < PT_3$, so only PT_3 can be 26 m, which implies that the treasure lies outside the triangle formed by the trees.

Teaching Suggestions:

For many problems, an accurate diagram shows the answer. With this problem, an extreme degree of precision is needed to show the relationship of the treasure's position to the triangle formed by the trees. A little trigonometry shows that angle T_3SP measures only slightly more than 60°.

$$\tan(\angle T_3SP) = \frac{26}{15}$$
$$= 1.7333\overline{3}$$
$$m(\angle T_3SP) \approx 60°01'$$

The key to solving this problem is for students to realize that the measure of $\angle T_3SP$ is greater than 60°.

EXTENSION FOR PROBLEM 15

ONE UP

A soft drink manufacturer delivers its product in cases holding exactly 40 bottles packed side by side. A note in the company suggestion box pointed out that the firm could pack 41 bottles in the same cases by simply rearranging the bottles.

The diagram at the left shows a full case of 40 bottles. The figure at the right is an empty case. Show how 41 bottles can be arranged in the empty case.

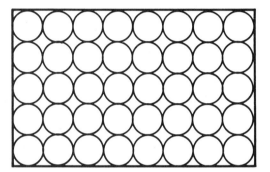 

Discussion for
ONE UP

Answer:

Make 5 five-bottle rows and 4 four-bottle rows. Alternate between five-bottle rows and four-bottle rows.

Clues:

1. In a side-by-side packing, the centers of the circles (representing the bottles) are _____.
2. What geometric shapes tessellate the plane?

Solution:

Assume that the radius of each bottle is 1 unit. Then the packing shown in the given diagram fills a case measuring 10 units × 16 units.

Note that the circles in the packing fit together to make up the 10 unit × 16 unit box. When the centers of the circles are connected, they form an array of squares. Call this a *square-array packing*.

By changing to a *triangular-array packing* (packing the circles so their connected centers form an array of triangles), more spaces can be created. Rows of circles will alternate between five-bottle rows and four-bottle rows. Each five-bottle row fills the 10-unit width.

The triangles formed by connecting centers of the circles are equilateral triangles measuring 2 units on a side. The height of these triangles represents the horizontal distance between rows. This distance is $\sqrt{3}$ units, so the total horizontal distance covered by this configuration is given by the following.

$$\binom{\text{number}}{\text{of rows}} \times \binom{\text{distance}}{\text{between rows}} + 2\binom{\text{distance from center}}{\text{of circle to edge}}$$
$$= (\ 8\ \ \times\ \sqrt{3})\ \ \ \ \ \ \ \ \ \ + 2(1)$$
$$\approx 15.86$$

Since the full length of the box is 16 units, the new triangular-array packing of bottles will easily fit into the box.

Teaching Suggestions:

You will find that modeling this problem using pennies or small equal-radii discs is very helpful for getting students started. Ask questions like, "How many bottles will fit into the case if the first row has four (rather than five) bottles, the second row has five bottles, the third has four, and so on?" Once students realize that they can deviate from a square-array packing and figure out the dimensions of the box, they will quickly find their way to a solution.

A sheet of graph paper that allows you to draw both tessellations accurately is included in the back of this book.

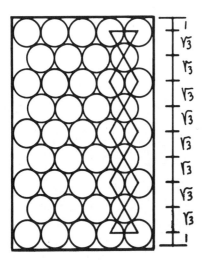

WARM-UP FOR PROBLEM 16

CANNONBALL RUN

Rufus Leaking stores his collection of cannonballs in cubical boxes that have no tops.

- The volume of each box equals its surface area. (Units are in cubic feet.)
- The volume of each cannonball equals its surface area. (Units are in cubic inches.)

How many cannonballs can Rufus fit into each box?

Discussion for
CANNONBALL RUN

Answer:

With the volume of each box given in cubic feet and the volume of each cannonball given in cubic inches, Rufus can fit at least 1000 cannonballs into a box. Rufus can fit in as many as 1254 balls if he uses the best possible stacking.

Clues:

1. Find the dimensions of a box and the radius of a cannonball.
2. The volume of a box is s^3 where s stands for the length of a side. The surface area of a box with no top is $5s^2$.
3. If r is the radius of a sphere, then its volume is $(4/3)\pi r^3$ and its surface area is $4\pi r^2$.
4. If your answer is 1000, you have not found the best way to stack the cannonballs.
5. Suppose Rufus can fit 100 cannonballs in the bottom of a box. How many balls can he fit into the next layer if each ball comes in contact with four balls from the bottom layer?
6. The bottom layer can actually hold either 104 or 105 cannonballs.

Solution:

Let s stand for the length of the side of a box. Then the volume of the box is s^3 and its surface area is $5s^2$. But these two values are equivalent, so the box must be $5 \times 5 \times 5$.

$$s^3 = 5s^2$$
$$s^3 - 5s^2 = 0$$
$$s^2(s - 5) = 0$$
$$s = 0 \text{ or } 5$$

Similarly, if r stands for the radius of a cannonball, then its volume is $(4/3)\pi r^3$ and its surface area is $4\pi r^2$, giving a radius of 3 units.

$$\frac{4}{3}\pi r^3 = 4\pi r^2$$
$$\frac{4}{3}\pi r^3 - 4\pi r^2 = 0$$
$$4\pi r^2 \left(\frac{1}{3}r - 1\right) = 0$$
$$r = 0 \text{ or } 3$$

At this point in the problem you must make some assumptions about the uniformity of the pieces involved and about the units in which the dimensions are given. For the purposes of this solution,

- assume that all the boxes are the same size;
- assume that all the cannonballs are the same size.

EACH LAYER PACKED SQUARELY, LAYERS PACKED SQUARELY

A quick solution to this problem, under the assumptions, is to divide the box dimensions by the value of the diameter of a cannonball.

(5 ft/box side) (12 in./ft) ÷ (6 in./ball) = 10 balls/box side

This calculation would tell you how many balls you could fit into the box each way; that is, you can fit 10 balls × 10 balls in the bottom of the box and you can make 10 layers of balls. In other words, Rufus can fit $10 \times 10 \times 10$ balls, or 1000 balls, in the box — 10 layers of 100 balls packed in a square-array configuration.

EACH LAYER PACKED SQUARELY, LAYERS SANDWICHED

A more efficient packing starts, as before, by placing 100 cannonballs (10 × 10) in the bottom of the box. The second layer is formed by placing balls in the indentation formed where four balls of the bottom layer come together.

A quick count shows that this second layer will have 81 cannonballs (9 × 9). Alternating layers of 100 balls with layers of 81 balls, Rufus will be able to make 6 layers of 100 balls and 6 layers of 81 balls before he reaches the top, giving him a total of 1086 cannonballs (6 · 100 + 6 · 81).

EACH LAYER IN TRIANGULAR ARRAY, LAYERS SANDWICHED

An even more efficient packing puts each layer of balls in a triangular array. The bottom layer can hold 6 rows of 10 cannonballs each alternating with 5 rows of 9 cannonballs each, giving a total of 105 balls (6 · 10 + 5 · 9).

The second layer is formed by placing balls in the indentation formed where three balls of the bottom layer meet.

CONTINUED ON PAGE 153

GEOMETRY PROBLEM 16

MAKING ARRANGEMENTS

Lester Turlbutt has packed spheres that are one-unit in diameter into a rectangular tray, filling the tray in a single layer with no slack, using a rectangular packing. But Lester wants to add more spheres. First he tries a different arrangement that allows him to add one sphere. But then, using a third arrangement, he finds that he is able to fit in still another sphere. What are the dimensions of the tray?

Discussion for
MAKING ARRANGEMENTS

Answer:

The tray's dimension are 6 units × 8 units or 5 units × 24 units.

Clues:

1. Use pennies to make a model of the problem.
2. What ways can you fill the tray other than by a rectangular (square-array) packing?
3. How long must the tray be in order to pick up an extra row for your alternate packings?
4. The alternate packings will have an odd number of rows.
5. For one solution, the dimensions of the tray are consecutive even integers.

Solution:

The solution to WARM-UP PROBLEM 16 showed a way to rearrange a packing to allow for an extra bottle. It changed the packing from a square array to a triangular array. The key to the solution was that the new array added extra rows.

If you work with a variety of small arrays of pennies, say 2 × 3 or 4 × 6, you will find that changing to a triangular-array packing does not allow you to add an extra row (adding rows is the only way you'll be able to add pennies using a triangular array); you don't pick up enough space. The first question to be answered, then, is, "At what point does a triangular-array packing give you an extra row?"

You will recall that the distance between the rows of a triangular-array packing was $r\sqrt{3}$, where r stands for the radius of the circle. In this case, then, the distance between rows is $(\sqrt{3}/2)$ units. Using this fact, you can quickly discover how many rows you need, namely 8.

NUMBER OF ROWS IN RECTANGULAR PACKING	LENGTH	LENGTH OF TRIANGULAR PACKING HAVING AN ADDITIONAL ROW
1	1	$2\left(\frac{1}{2}\right) + 1\left(\frac{\sqrt{3}}{2}\right)$, or about 1.87
2	2	$2\left(\frac{1}{2}\right) + 2\left(\frac{\sqrt{3}}{2}\right)$, or about 2.73
3	3	$2\left(\frac{1}{2}\right) + 3\left(\frac{\sqrt{3}}{2}\right)$, or about 3.60
4	4	$2\left(\frac{1}{2}\right) + 4\left(\frac{\sqrt{3}}{2}\right)$, or about 4.46
5	5	$2\left(\frac{1}{2}\right) + 5\left(\frac{\sqrt{3}}{2}\right)$, or about 5.33
6	6	$2\left(\frac{1}{2}\right) + 6\left(\frac{\sqrt{3}}{2}\right)$, or about 6.20
7	7	$2\left(\frac{1}{2}\right) + 7\left(\frac{\sqrt{3}}{2}\right)$, or about 7.06
8	8	$2\left(\frac{1}{2}\right) + 8\left(\frac{\sqrt{3}}{2}\right)$, or about 7.93

Next, investigate the columns. First notice that no matter how many columns you start with, when you create a triangular array from a packing that has eight rows, you will make a packing that has five rows of one number and four rows of the other. For example, rearranging a rectangular packing that has 8 rows of two spheres each will yield either five rows with two spheres each alternating four rows with one sphere each or it will yield five rows with one sphere each alternating four rows with two spheres each. The following table shows the results as you systematically add columns to the possible rectangular packing.

NUMBER OF COLUMNS IN RECTANGULAR PACKING	NUMBER OF SPHERES	NUMBER OF SPHERES IN TRIANGULAR PACKINGS
2	16	13 or 14
3	24	22 or 23
4	32	31 or 32
5	40	40 or 41
6	48	49 or 50

Not until you reach six columns in the rectangular packing will you be able to meet the conditions of Lester's arrangements.

CONTINUED ON PAGE 154

EXTENSION FOR PROBLEM 16

SQUARE UNITS

The following diagram shows an arrangement of squares that appeared on the cover of the November 1958 issue of *Scientific American*. If the area of square C is 64 square units and the area of square D is 81 square units, what are the areas of the other seven squares?

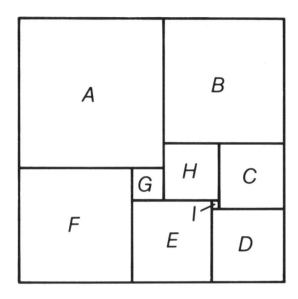

Discussion for

SQUARE UNITS

Answer:

The areas are as follows.

A is 324 units^2, B is 225 units^2, C is 64 units^2, D is 81 units^2, E is 100 units^2, F is 196 units^2, G is 16 units^2, H is 49 units^2, I is 1 unit^2

Clues:

1. The difference between the greatest area and the least area is 323 units^2.
2. No two squares have the same area.
3. The complete figure is not a square. It is 33 units wide and 32 units high.
4. The area of each square is an integer.

Solution:

Since square D is 81 units^2 in area, its dimensions are 9 units $\times$ 9 units. Similarly, square C, with an area of 64 units^2, is an 8 unit $\times$ 8 unit square. Square I, then, must be a 1 unit $\times$ 1 unit square and square E, in turn, must be a 10 unit $\times$ 10 unit square (since square E's side is the length of square I's side plus square D's side) and square H must be a 7 unit $\times$ 7 unit square (square C's side less square I's side).

The sides of squares G and H add to 11 units ($E + I$), so square G must be 4 units $\times$ 4 units ($11 - H = 11 - 7 = 4$).

One side of square F is partitioned by squares G and E, so square F is a 14 unit $\times$ 14 unit square ($G + E = 4 + 10 = 14$). Square A has a side made up of square F and square G, so it measures 18 units $\times$ 18 units ($F + G = 14 + 4 = 18$). Finally, the sides of squares H and C add to 15 units ($7 + 8 = 15$), so square B must be a 15 unit $\times$ 15 unit square.

The answer is found by squaring the dimensions of each square.

Teaching Suggestions:

The lengths of the sides of the squares are easier to work with than the areas. Once that realization hits, students will have very little trouble with the problem. Simple deduction brings out the answers nicely.

WARM-UP FOR PROBLEM 17

A PERFECT FIT

Show how to make a large rectangle by fitting together these pieces.

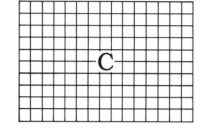
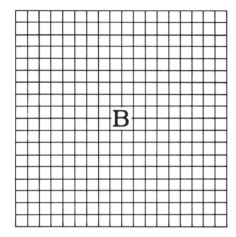
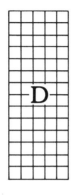
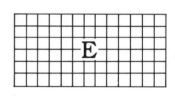
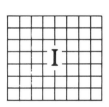
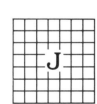
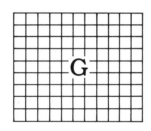
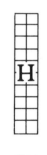
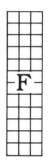

Discussion for

A PERFECT FIT

Answer: See diagram in Solution.

Clues:

1. What is the total area of the eleven pieces?
2. What are the only possible dimensions of the large rectangle you're trying to make?
3. Which piece or pieces must lie along the length of the rectangle?
4. On a piece of graph paper, draw the large rectangle. Cut out replicas of the eleven pieces to fit into your rectangle.
5. If the long side of A were to lie along the longest side of the rectangle, which piece could be used to fill in the rest of that space?

Solution:

The dimensions and areas of the eleven pieces are given in the following table.

PIECE	DIMENSIONS	AREA
A	25 × 4	100
B	18 × 18	324
C	15 × 10	150
D	5 × 14	70
E	13 × 6	78
F	3 × 12	36
G	11 × 9	99
H	2 × 10	20
I	8 × 7	56
J	7 × 7	49
K	2 × 2	4

The area of the eleven pieces together is 986 units^2. The prime factors of 986 are 2, 17, and 29, so the only possible dimensions for the rectangle you're trying to make are 2 × 493, 17 × 58, or 29 × 34.

If A were to lie along a 29 unit edge, only 4 units would remain for other pieces. The only pieces left to fill that gap are H and K. (F cannot be used since no piece would be left to fill the remaining 1 unit space.) But K is too small; it creates a 2 × 2 gap that cannot be filled. Therefore, the long side of A must align with the 34 unit side of the large rectangle.

B is an 18 × 18 piece, so it will take up 18 units of the large rectangle from both dimensions. There are only 7 units to be filled along the 29 unit edge (29 − 18 − 4) in some places and 11 units in others. Along the 34 unit edge there are either 16 units or 9 units to be filled. Consider piece C, which is 10 × 15. If the longest side of C lay along the 34 unit side, only a 1 unit gap would remain. This gap can't be filled, so the long side of C must align with the 29 unit side.

In a similar fashion, working from the largest pieces to the smallest pieces, each piece will locate itself within the rectangle. The following diagram gives their positions.

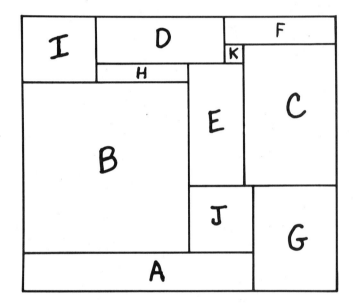

Teaching Suggestions:

Working with cut-out pieces makes this problem much more manageable. You might suggest that students keep all their cut-out pieces in an envelope so they won't be lost.

The solution analyzes the placement of the large pieces first. There are fewer options for these pieces than for the small pieces. Also, proceeding in such a systematic fashion makes it easier to keep track of whatever solutions have been tried and eliminates some of the guesswork inherent in this problem.

GEOMETRY PROBLEM 17

CRAYONA'S CANVAS

Crayona Lustoire, the eccentric artist, claims that her best canvasses have the same area and perimeter (that is, the number of units of area equals the number of units of perimeter). If the unit measures are integers, what are the dimensions of Crayona's canvasses?

Discussion for

CRAYONA'S CANVAS

Answer:

Crayona's canvasses are either 4 units × 4 units or 3 units × 6 units.

Clues:

1. This problem has more than one answer.
2. One kind of canvas is square-shaped.
3. Write an equation describing the equality of area and perimeter for a rectangle. (How could your equation be simplified for a square?)
4. There are only two such canvasses.

Solution:

Let x stand for one dimension of a canvas and y stand for the other dimension. Then the area of the canvas is xy and its perimeter is $2x + 2y$. These two values are equal.

$$xy = 2x + 2y$$
$$x(y - 2) = 2y$$
$$x = \frac{2y}{y - 2}$$

A quick way to find appropriate integer values for the dimensions is to substitute various integer values for y and calculate x.

y	x
1	−2
2	undefined
3	6
4	4
5	3.3
6	3
7	2.8
8	2.6
9	2.57
10	2.5
20	2.$\overline{2}$
100	2.04
1000	2.004

The chart shows that dimensions of 3 × 6 and 4 × 4 are acceptable. It also suggests that, for no values of y beyond 6, will there be any more acceptable values for x. The values of x will all be between 2 and 3 (as values of y increase, values of x approach 2).

There is a nice geometric argument for why 4 × 4 and 3 × 6 are the only acceptable dimensions. Take any rectangle with integer dimensions. Separate the figure into unit squares.

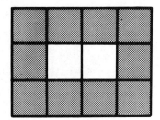

Now, look at the border squares (shaded). The perimeter is twice the number of squares along one horizontal side plus twice the number of squares along one vertical side. You can find the perimeter by counting all the border squares, including all four corners twice. That is, the perimeter of the figure is the sum of the number of border squares *plus four*.

The area of the figure is the number of unit squares that make it up. That is, the area of the figure is the sum of the number of border squares *plus the number of nonborder squares*.

In order for the area of the figure to equal the perimeter of the figure, the number of nonborder squares must equal four. That is, the desired rectangles must consist of a set of border squares and exactly four inner, nonborder squares. And there are only two configurations of squares that give that result, since there are only two possible ways to rearrange four squares.

Teaching Suggestions:

Many of my students have difficulty really accepting the geometric solution as adequate proof for only two possibilities. The numerical argument based on the table may help.

Graph paper is very useful for charting out different results.

EXTENSION FOR PROBLEM 17

TIGHTENING YOUR BELTS

Suppose a belt tightly stretched about the equator of the earth just fits. How long a piece should be inserted so that the belt could encircle the earth at a distance of 1 m away from the equator at all points?

Discussion for

TIGHTENING YOUR BELTS

Answer: 2π meters, or about 6.28 m

Clues:

1. The radius of the earth is about 6440 km.
2. The radius of the new belt is $(6440 \times 1000) + 1$.
3. Suppose you don't know the radius of the earth. How would you solve this problem?
4. If r represents the radius of the earth at the equator, what represents the radius of the extended belt?

Solution:

Assume the equator is a circle. Let r stand for the radius of the equator in meters. The circumference of the equator is $2\pi r$ meters. The circumference of a belt 1 m away from the equator at all points is $2\pi(r + 1)$ meters.

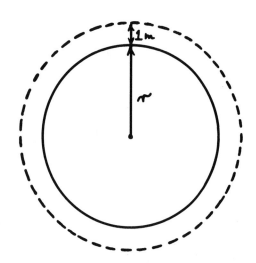

To find the length of the insert, subtract.

$$2\pi(r + 1) - 2\pi r = 2\pi$$

That is, the insert is 2π meters long, or about 6.28 m.

Teaching Suggestions:

There is no reason why students must solve this problem using variables. They can equally well use the real figures. The generalized solution is valuable to know about and understand, though. It shows that there are sometimes ways to solve problems in the absence of hard data and to avoid "hairy" calculations. For students having difficulties with generalizations, create a table listing various values for radii and their corresponding circumferences. Then find the circumferences for the larger circle and have students calculate the distances.

You can demonstrate this problem on a chalkboard by using a board compass and two pieces of string. Use the string to lay off line segments equivalent in length to the circumferences of the two circles.

WARM-UP FOR PROBLEM 18

HEAD START

The track coach at Aardvark High plans to make 8 lanes each 1 meter wide on the running track.

- The track measures *y* meters along the inside curb.
- The track has straight parallel sides and semicircular ends.
- If the runners in a race lined up at the same spot and stayed in their own lanes throughout the race, they would *not* run the same distance.

To make a race even, by how much should each lane be staggered?

Discussion for

HEAD START

Answer:

The lanes differ by 2π m. That is, since the innermost lane is y meters long, the next inside lane should be $(y + 2\pi)$ meters long, the next $(y + 4\pi)$ meters long, and so on up to $(y + 14\pi)$ meters.

Clues:

1. Draw a diagram of the track.
2. What changes affect the distances run by the different athletes.
3. Let r stand for the radius of the semicircular ends of the inside curb. What are the radii of the inside curves of each lane, given in terms of r?
4. Find the difference in distances between two consecutive lanes.
5. Look for a pattern.

Solution:

The following diagram shows the lanes of the track, numbered 1–8. The length of each straight parallel side is w meters and the radius of each semicircular end is r meters.

Since the lanes are 1 m wide, the radii of the inside of the semicircular parts of each lane increase by 1 m. That is, the radius for Lane 1 is r meters, the radius for Lane 2 is $(r + 1)$ meters, the radius for Lane 3 is $(r + 2)$ meters, ..., the radius for Lane 8 is $(r + 7)$ meters. The straightaways (along the parallel sides) for each lane are the same length. The lengths of each lane are as follows.

INSIDE LANE 1 $= 2\pi r + 2w$
INSIDE LANE 2 $= 2\pi(r + 1) + 2w = (2\pi r + 2w) + 2\pi$
INSIDE LANE 3 $= 2\pi(r + 2) + 2w = (2\pi r + 2w) + 4\pi$
INSIDE LANE 4 $= 2\pi(r + 3) + 2w = (2\pi r + 2w) + 6\pi$
INSIDE LANE 5 $= 2\pi(r + 4) + 2w = (2\pi r + 2w) + 8\pi$
INSIDE LANE 6 $= 2\pi(r + 5) + 2w = (2\pi r + 2w) + 10\pi$
INSIDE LANE 7 $= 2\pi(r + 6) + 2w = (2\pi r + 2w) + 12\pi$
INSIDE LANE 8 $= 2\pi(r + 7) + 2w = (2\pi r + 2w) + 14\pi$

The lanes differ by 2π meters.

Teaching Suggestions:

Although students may be happier working with actual distance, the solution is independent of y. If your students have trouble generalizing formulas, you might start them out by giving them a value for the radius of the inside semicircle and having them work from there. Understanding that the only changes in distance traveled are along the semicircular ends is the key to solving this problem.

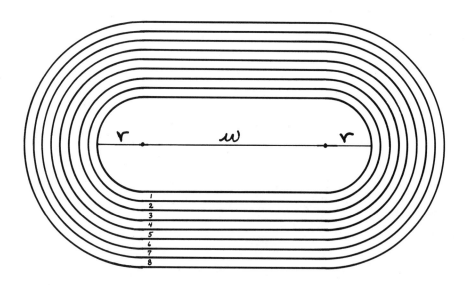

GEOMETRY PROBLEM 18

HIGH ROLLERS

The Egyptians used rollers to move the big blocks that went into the pyramids. Suppose a block is supported on two rollers, each 21 cm in diameter. How far can the block advance in one complete revolution of the rollers?

Discussion for

HIGH ROLLERS

Answer: 42π centimeters, or about 131.88 cm

Clues:

1. The circumference of a circle is given by $C = \pi d$.
2. If the rollers turned in place, how far would the block move?
3. The rollers are moving forward.

Solution:

If the rollers simply revolved in place (like a conveyor belt), they would move the block forward the length of a circumference, 21π. But the rollers are moving along the ground at the same time—moving the distance of their circumferences; that is, another 21π centimeters.

Any number of rollers placed under the block and moving together will produce the same advancement.

Teaching Suggestions:

The rollers *moving* along the ground make an important contribution to the answer of this problem. A simple model can be made with paper towel rollers and a ruler. Mark a spot on the circumference of a roll, place the ruler on the roller, and roll the ruler forward one revolution. Then analyze the results.

EXTENSION FOR PROBLEM 18

AT ROPE'S END

A cow is tethered by a piece of rope 50 m long. The rope is fastened to a hook located 10 m from the corner of the longest side of a barn measuring 60 m × 30 m. Over how much ground can the cow graze?

Discussion for
AT ROPE'S END

Answer: 1675π m², or about 5259.5 m²

Clues:

1. Draw a sketch of the problem.
2. If r stands for the radius, the area of a semicircle is _____.
3. If r stands for the radius, the area of a quarter-circle is _____.
4. The total area is the sum of the areas of a semicircle and two different quarter-circles.
5. What are the radii of the two quarter-circles?

Solution:

The diagram shows the areas the cow can cover.

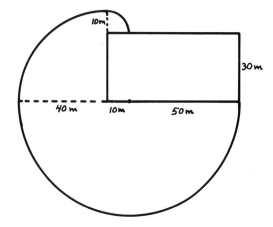

As the diagram suggests, the cow's grazing area is the area of a semicircular region with a 50 m radius plus the area of a quarter-circular region with a 40 m radius plus the area of a quarter-circular region with a 10 m radius.

$$\text{total area} = \frac{1}{2}\pi(50)^2 + \frac{1}{4}\pi(40)^2 + \frac{1}{4}\pi(10)^2$$

$$= \frac{1}{4}\pi[2 \cdot 50^2 + 40^2 + 10^2]$$

$$= 1675\pi$$

$$\approx 5259.5$$

The cow can graze over approximately 5259.5 m² of ground.

Teaching Suggestions:

Encourage students to sketch the problem and identify each part of the total area. Once they have drawn an accurate sketch, they should be able to proceed on their own.

This problem deserves a few good follow-up questions such as the following.

- Would the grazing area remain the same if the hook were located 10 m from the corner of the shorter side of the barn? Greater or smaller? By how much?
- Would the grazing area change if the hook were located on a corner? If so, how would it change and by how much?

WARM-UP FOR PROBLEM 19

PARTIAL ECLIPSE

What is the exact area of the shaded region?

- The figure consists of a partial circle and a semicircle extended by two line segments.
- The width of the figure at its widest point is the same as the diameter of the circle, 110 m.
- The width of the partial circle is 22 m less than the diameter of the circle.

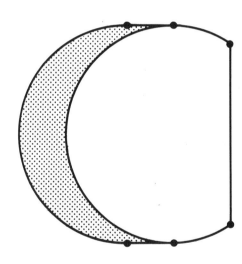

Discussion for

PARTIAL ECLIPSE

Answer: 2420 m²

Clues:

1. To find the area of the shaded region, you can find the area of another figure hidden in the diagram.
2. What kind of figure is figure *AEGC*?

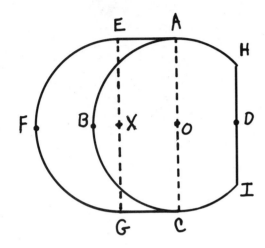

3. Don't worry about figure *ACIH*; you don't need it.
4. What are two different ways (using addition) to represent the area of figure *AEFGC*?
5. Area (*AEFGC*) = Area (*AEGC*) + Area (?)
   Area (*AEFGC*) = Area (*ABC*) + Area (?)
6. How do the areas of semicircles *ABC* and *EFG* compare?

Solution:

The shaded crescent has the same area as the rectangle formed by joining *A*, *C*, *E*, and *G*.

There are two ways to express the area of region *AEFGC*:

1. Area of semicircle *EFG* + Area of rectangle *AEGC*
2. Area of semicircle *ABC* + Area of crescent *AEFGC*

Because these two expressions represent the same area, they are equal to one another. Semicircle *EFG* and semicircle *ABC* have the same diameter so the areas must be the same, leaving the area of the rectangle and the area of the crescent the same also. The area of the rectangle is 110 × 22 meters or 2,420 m².

Teaching Suggestions:

The techniques used in this problem—focusing on the simple geometric parts of a complex figure—are extremely helpful in a variety of geometric problems. Many of my students groan when they are faced with this problem. They think they must find the area of the missing part of the circle. Once they realize there are ways around the calculation, they are much more enthusiastic about the problem. Discovering the usefulness of the rectangle is the most important part of this problem.

You may wish to point out that crescent *FEABCG* is not a true crescent since it is not made entirely from arcs of circles. The term crescent is used in the solution only for ease of identification.

IN THE SHADE

Inside this quarter-circle are two semicircles having the same radius, r. Which of the two shaded regions has the greater area?

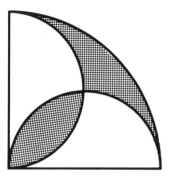

Discussion for

IN THE SHADE

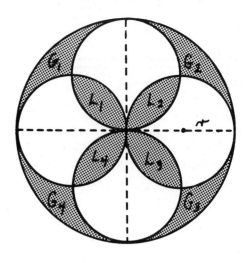

Answer: The areas are the same.

Clues:

1. Draw the entire figure, completing the quarter-circle and four circles derived from the semicircles.
2. What is the radius of the large circle?
3. How do the areas of the four small circles compare to the area of the large circle?
4. Write an expression to represent the total area of the outer shaded regions.
5. Write an expression to represent the total area of the inner shaded regions.

Solution:

The diagram shows the entire figure with all the circles completed. G_1, G_2, G_3, and G_4 are names for the outer shaded regions (ginko leaves); L_1, L_2, L_3, L_4 are names for the inner shaded regions (lenses). The small circles each have the same radius, r. The radius of the large circle equals the diameter of the small circles, $2r$.

The areas can be described by the following series of equations.

$$\text{Area}(G_1) = \frac{1}{4}[\text{Area}(G_1 + G_2 + G_3 + G_4)]$$

$$= \frac{1}{4}[\text{Area (large circle)} - \text{Area (small overlapping circles)}]$$

$$= \frac{1}{4}[\pi(2r)^2 - \text{Area (small overlapping circles)}]$$

$$= \frac{1}{4}[4\pi r^2 - \text{Area (small overlapping circles)}]$$

$$= \frac{1}{4}[\text{Area (small circles not overlapping)} - \text{Area (small overlapping circle)}]$$

$$= \frac{1}{4}[\text{Area}(L_1 + L_2 + L_3 + L_4)]$$

$$= \text{Area } L_1$$

In other words, the areas of the two shaded regions, G_1 and L_1, are equal.

Teaching Suggestions:

Encourage students to represent the combined areas in different ways. It helps them to arrive at the correct answer. A problem I sometimes use at this point involves a large circle with four smaller circles inside, each tangent to the large circle and tangent to the two adjacent circles also. I ask students to find the radius of the smaller circles. (It's $3(\sqrt{2} - 1)$).

EXTENSION FOR PROBLEM 19

SEMI-SQUARE

If the area of a square inscribed in a circle is 15 cm², what is the area of the square inscribed in a semicircle of the same circle?

Discussion for

SEMI-SQUARE

Answer: 6 cm²

Clues:

1. Draw a diagram. Label the vertices of the large square A, B, C, and D. Label the corresponding vertices of the small square A', B', C', and D'. Label the center of the circle O.
2. Find a value for the radius of the circle.
3. Suppose r stands for the radius of the circle. What is the length of a side of the large square in terms of r? The small square?

Solution:

In the diagram below, O is the center of the circle, figure ABCD is the square with the 15 cm² area, figure A'B'C'D' is the square inscribed in the semicircle.

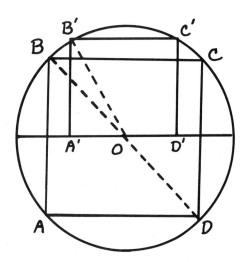

Let r stand for the radius of the circle. Then $BD = 2r$. Using the Pythagorean Theorem, you can find a value for r^2 since the area of square ABCD equals AB^2.

$$AB^2 + AD^2 = BD^2$$
$$2(AB)^2 = BD^2$$
$$2(15) = (2r)^2$$
$$30 = 4r^2$$
$$\frac{15}{2} = r^2$$

In the small square, $A'O = (1/2)A'D' = (1/2)A'B'$ and the area of the small square equals $(A'B')^2$. Using the Pythagorean Theorem again, you can find a value for the area of the small square.

$$(A'B')^2 + (A'O)^2 = (OB')^2$$
$$A'B'^2 + \left(\frac{1}{2}A'B'\right)^2 = r^2$$
$$\frac{5}{4}(A'B')^2 = r^2$$
$$(A'B')^2 = \frac{4}{5}r^2$$
$$= \frac{4}{5}\left(\frac{15}{2}\right)$$
$$= 6$$
Area $(A'B'C'D') = 6$

The area of the small square is 6 cm².

Teaching Suggestions:

Once students are able to see the relationship between the radii and the sides of the squares, they are generally able to proceed on their own.

WARM-UP FOR PROBLEM 20

SEEMORE'S SYMBOL

The official symbol of the *Society of Mathematical Enthusiasts* at Seemore University is a circle with an equilateral triangle inscribed and another circumscribed about the circle. The difference between the areas of the two triangles is 25 cm². What is the radius of the circle?

Discussion for
SEEMORE'S SYMBOL

Answer: $\dfrac{10}{3}\sqrt{\dfrac{\sqrt{3}}{3}}$ centimeters, or about 2.533 cm

Clues:

1. What is the formula for the area of an equilateral triangle in terms of the length of its sides?
2. Find values for the lengths of the sides of the two triangles expressed in terms of the value of the radius.
3. What are the relationships among the sides of a 30°-60° right triangle?

Solution:

In the following diagram, $\triangle ABC$ and $\triangle DEF$ are the equilateral triangles and the given circle has center O.

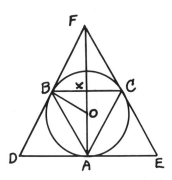

Because the two triangles are regular, O is their center. Also $\overline{OA} \cong \overline{DE}$ and $\overline{OB} \cong \overline{DF}$ and $\overline{OC} \cong \overline{FE}$. So, because $\triangle DEF$ is equilateral, points A, B, and C bisect the sides of $\triangle DEF$. From this information, you can conclude that $\triangle ABD$ is equilateral ($AD = DB$ and $m\angle D = 60°$). Similarly, $\triangle ACE$ and $\triangle BCF$ are equilateral and congruent to $\triangle ABD$ and $\triangle ABC$, and, from this, you can conclude that the length of a side of the large triangle, $\triangle DEF$, is twice the length of a side of the small triangle, $\triangle ABC$.

To solve the problem, follow this procedure.

- Find the area of the small triangle, $\triangle ABC$, expressed in terms of the circle's radius, r.
- Find the area of the large triangle, $\triangle DEF$, expressed in terms of the circle's radius, r.
- Find the difference of the area expressions, equate it to 25, and solve the equation for r.

Look at $\triangle BXO$. It's a 30°-60° right triangle. (Why?) The right angle is at X, the 30° angle is at B, and the 60° angle is at O. The length of $\overline{BO}$ is r, so $OX = r/2$ and $BX = (r\sqrt{3})/2$. But BX is half the length of BC, so $BC = r\sqrt{3}$. The area of $\triangle ABC$ can be found as follows.

$$\text{Area}(\triangle ABC) = \frac{1}{2}(\text{base})(\text{height})$$
$$= \frac{1}{2}(BC)(AO + OX)$$
$$= \frac{1}{2}(r\sqrt{3})\left(r + \frac{r}{2}\right)$$
$$= \frac{3r^2\sqrt{3}}{4}$$

The area of $\triangle DEF$ is four times that of $\triangle ABC$, so Area $(\triangle DEF) = 3r^2\sqrt{3}$.

Since the difference between the two areas is 25 cm², you can complete the problem as follows.

$$\text{Area}(\triangle DEF) - \text{Area}(\triangle ABC) = 25$$
$$3r^2\sqrt{3} - \frac{3r^2\sqrt{3}}{4} = 25$$
$$12r^2\sqrt{3} - 3r^2\sqrt{3} = 100$$
$$r^2(9\sqrt{3}) = 100$$
$$r = \frac{10}{3}\sqrt{\frac{\sqrt{3}}{3}}$$
$$\approx 2.533$$

The radius of the circle is about 2.533 cm.

Teaching Suggestions:

Once students find a way to represent the sides of the triangles in terms of the circle's radius, they will be able to proceed in a straightforward manner.

To extend the problem, I sometimes ask students to find the area of the circle or the area of the three outer segments of the circle.

INTEGRALLY YOURS

The sides of a triangle measured in centimeters are integer units. The area in square centimeters is also in integer units. Find the length of the shortest side of the triangle if one side is 21 cm and the perimeter is 48 cm.

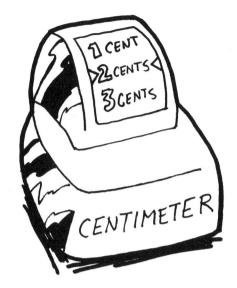

Discussion for
INTEGRALLY YOURS

Answer: 10 cm

Clues:

1. The shortest side is *not* 21 cm long.
2. Let x represent the length of the shortest side. Express the length of the third side in terms of x.
3. Sketch the triangle. Draw an altitude to the 21 cm side.
4. What is a Diophantine equation? What do Diophantine equations have to do with this problem?

Solution:

Let x stand for the length of one side of the triangle. Since another side of the triangle is 21 cm long and the perimeter is 48 cm, the length of the third side is given by $27 - x$. Either x or $27 - x$ represents the shortest side, for, if x were greater than 21, then $27 - x$ would be less than 21. If you assume that x represents the shortest side, then the only possible values of x and $27 - x$ are given in the following table. (They must be integer values.)

x	$27 - x$
1	26
2	25
3	24
4	23
5	22
6	21
7	20
8	19
9	18
10	17
11	16
12	15
13	14

The following diagram shows the triangle with an altitude drawn to the 21 cm side. The altitude, with length a, separates the side into two parts; their lengths are represented by y and $21 - y$.

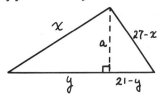

By using the Pythagorean Theorem, you can express the square of the altitude in two different ways.

$$a^2 = x^2 - y^2 \text{ and } a^2 = (27 - x)^2 - (21 - y)^2$$

Equating these two expressions gives a value for x in terms of y.

$$x^2 - y^2 = (27 - x)^2 - (21 - y)^2$$
$$x^2 - y^2 = (729 - 54x + x^2) - (441 - 42y + y^2)$$
$$54x = 288 + 42y$$
$$x = \frac{48 + 7y}{9}$$

Since x must be an integer, only certain values of y are permissible—those that make the expression $(48 + 7y)/9$ an integer. The table below suggests what values of x are possible, namely 6, 7, 8, 9, 10, 11, 12, and 13.

$x = \dfrac{48 + 7y}{9}$	y
5⅓	0
6	6/7
6⅑	1
6⁸⁄₉	2
7	2 1/7
8	3 3/7
9	4 5/7
10	6
11	7 2/7
12	8 4/7
13	9 6/7

The next question is, "Which of these values will assure an integer value for the area of the triangle?" The area of the triangle is given by the following.

$$\text{Area (triangle)} = \frac{1}{2}(21)(a)$$
$$= \frac{1}{2}(21)\sqrt{x^2 - y^2}$$

The values of x and y must be such that not only is $x^2 - y^2$ an integer, but an *even perfect square*. For such conditions to hold, y must be an integer since x is an integer. So, the only possible value for x is 10.

Teaching Suggestions:

Most students will readily accept the answer 10, but will not entirely understand why the other values in the second chart are not acceptable. You might have them calculate $x^2 - y^2$ (or the area) for one or two pairs of values. They should begin to see that y must be an integer.

I have found that this problem allows me to review slopes, intercepts, graphing, and other concepts from first-year algebra.

TURN DOWN

The base of a triangular sheet of paper is 12 cm long. The paper is folded down over the base—the crease parallel to the base. The area of the part of the triangle projecting below the base is 0.36 times the area of the entire triangle. Find the length of the crease.

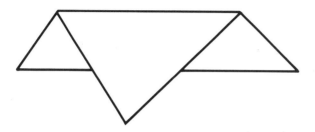

Discussion for
TURN DOWN

Answer: 9.6 cm

Clues:

1. Draw a picture of the triangle unfolded. Include broken lines to show the crease and to show where the folded triangle overlaps the base of the triangle.
2. What are the similar triangles in the figure?
3. Express 0.36 as a fraction.
4. How do the ratios of sides of similar triangles compare to the ratios of areas?
5. What kind of figure is formed by the triangle minus the projecting triangle?

Solution:

In the following diagram, $\triangle ABC$ represents the triangular sheet of paper, $\overline{DE}$ is the crease, and $\triangle MNC$ is the part of the triangle that extends below the base with $\overline{MN}$ the segment that lines up with the base.

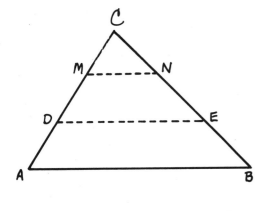

Since $\overline{MN}$ lines up with the base, $\overline{AB}$, the two line segments are parallel. Therefore, $\angle CMN \cong \angle CAB$ and $\angle CNM \cong \angle CBA$ because they are corresponding angles. You can conclude by angle-angle that $\triangle CMN \sim \triangle CAB$.

In similar triangles, the ratio of areas is directly proportional to the square of the ratio of corresponding sides. You can use this fact to find a value for MN.

$$\left(\frac{MN}{AB}\right)^2 = \frac{36}{100}$$

$$\left(\frac{MN}{12}\right)^2 = \frac{36}{100}$$

$$MN^2 = 12^2\left(\frac{36}{100}\right)$$

$$MN = 12\left(\frac{6}{10}\right)$$

$$= 7.2$$

Since $\overline{MN}$ lines up with $\overline{AB}$, the distance between $\overline{DE}$ and $\overline{MN}$ equals the distance between $\overline{DE}$ and $\overline{AB}$. In other words, $\overline{DE}$ is the median of trapezoid ABNM.

$$DE = \frac{1}{2}(MN + AB)$$

$$= \frac{1}{2}(7.2 + 12)$$

$$= 9.6$$

Teaching Suggestions:

A paper model makes this problem much easier to follow. Folding the projection over the base creates a second crease corresponding to $\overline{MN}$ in the diagram. The creases show the similar triangles clearly.

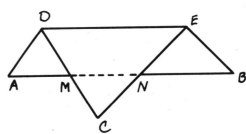

WARM-UP FOR PROBLEM 21

TRIANGLE PARK

On C. T. Planner's blueprint, Triangle Park measures 12 in. × 17 in. × 25 in. C. T. knows that the actual area of the park is 156,150 ft². What scale was used for the blueprint?

Discussion for

TRIANGLE PARK

Answer: 1 in. = 41 ft 8 in., or 12 in. = 500 ft

Clues:

1. How do the two triangles, Triangle Park and its blueprint, compare?
2. How do the areas of similar triangles compare?
3. The triangle is *not* a right triangle.
4. Let *Hero* do it for you.
5. Find the area of the blueprint triangle.

Solution:

In similar triangles, the ratio of areas is directly proportional to the square of the ratio of corresponding sides. The problem, then, is to find the area of the blueprint triangle.

Since the measures of all three sides of the triangle are given, you can use Hero's Formula to find the area. According to the formula, if the measures of the three sides are given by a, b, and c, then

$$\text{Area} = \sqrt{s(s-a)(s-b)(s-c)}$$

where $s = \frac{1}{2}(a + b + c)$

Convert the dimensions of the blueprint to feet. Then $a = 1$, $b = 17/12$, $c = 25/12$, and $s = 27/12$.

$$\text{Area}\begin{pmatrix}\text{blueprint}\\\text{triangle}\end{pmatrix} = \sqrt{\left(\frac{27}{12}\right)\left(\frac{15}{12}\right)\left(\frac{10}{12}\right)\left(\frac{2}{12}\right)}$$

$$= \frac{5}{8}$$

So the area of the blueprint triangle is 5/8 ft.²

The ratio of a side of the blueprint triangle corresponding to a side of the real park gives the scale of the blueprint. So, if a stands for a side on the blueprint triangle and a' is its corresponding side in the park, the scale can be found by the following calculation.

$$\left(\frac{a}{a'}\right)^2 = \frac{\text{Area (blueprint triangle)}}{\text{Area (Triangle Park)}}$$

$$= \frac{5/8}{156{,}250}$$

$$= \frac{1}{250{,}000}$$

$$\frac{a}{a'} = \frac{1}{500}$$

$$a' = 500$$

The scale is 1 ft = 500 ft, or 1 in. = 41 ft 8 in.

Teaching Suggestions:

I use this problem to review simplifying radicals. Encourage students to work with values in fractional form. The numbers work out nicely that way.

Hero's Formula is very convenient for problems like this. You could, however, find the area of the blueprint triangle by first using the Pythagorean Theorem to calculate an altitude and, then, proceeding via the standard area formula.

GEOMETRY PROBLEM 21

WRITER'S RECTANGLE

A park named Writer's Rectangle recently opened in our town. When asked about the dimensions of the rectangle, Jarvis (Gyp) Jameson, the city planner, responded with these clues:

- The diagonals of the rectangular park plus its longer sides together measure seven times one of the shorter sides.
- The length of one diagonal is 250 m longer than one of the shorter sides.

Use this information to find the area of the park.

Discussion for
WRITER'S RECTANGLE

Answer: 126,000 m²

Clues:

1. What kinds of triangles are involved?
2. Write an equation for each piece of information.
3. What is the ratio of the length of the diagonal to the length of the longer side?
4. Use substitution.
5. The answer to Clue 3 is 53:45.

Solution:

Let m stand for the length of the longest sides of the park, n stand for the length of the shortest sides of the park, and p stand for the length of the diagonals. Then you can write the following three equations in three unknowns.

FROM PYTHAGOREAN THEOREM: $n^2 = p^2 - m^2$
FROM FIRST CLUE: $7n = 2p + 2m$
FROM SECOND CLUE: $n = p - 250$

The problem asks for the area of the park. In terms of the variables, the area is given by $A = mn$, so the goal is to eliminate p and find a value for mn.

First, square both sides of the second equation.

$$49n^2 = 4(p + m)(p + m)$$

Then make a new equation using this equation and the equation derived from the Pythagorean Theorem.

$$49(p^2 - m^2) = 4(p + m)(p + m)$$
$$49(p - m) = 4(p + m)$$
$$49p - 49m = 4p + 4m$$
$$45p = 53m$$
$$p = \frac{53m}{45}$$

Next, use substitution.

$$7n = 2\left(\frac{53m}{45}\right) + 2m$$
$$315n = 106m + 90m$$
$$315n = 196m$$
$$n = \frac{28m}{45}$$

This information tells you that the ratios of the sides of the triangle formed by the diagonal and two sides are as follows.

$$n:p:m = 28:53:45$$

Now you can find values for the two sides and, hence, the area.

$$\frac{n}{p} = \frac{n}{n + 250}$$
$$\frac{28}{53} = \frac{n}{n + 250}$$
$$28n + 7000 = 53n$$
$$25n = 7000$$
$$n = 280$$

$$n = \frac{28m}{45}$$
$$280 = \frac{28m}{45}$$
$$m = \frac{280 \cdot 45}{28}$$
$$= 450$$

Area $= mn$
$= (450)(280)$
$= 126,000$

Teaching Suggestions:

This problem is a nice one for allowing you to review simultaneous equations. You may want to start students out on some simpler simultaneous equations first.

EXTENSION FOR PROBLEM 21

SMALLEY'S FOLLY

Jo Smalley never does things the easy way. When he wants to know the dimensions of a rectangular box, he asks for the areas of the top, side, and end. If those areas are 120 square units, 96 square units, and 80 square units, respectively, what are the dimensions of the box?

Discussion for

SMALLEY'S FOLLY

Answer: 12 units × 10 units × 8 units

Clues:

1. How many variables do you need to solve the problem?
2. Write three equations that describe the three given areas.
3. Use substitution to eliminate variables.
4. The dimensions are integer values.
5. You could solve this problem by examining prime factors.

Solution:

Let ℓ, w, and h stand for the length, width, and height of the box, respectively. Then you can write the following equations.

TOP: $\ell w = 120$
SIDE: $\ell h = 96$
END: $wh = 80$

SOLUTION BY SUBSTITUTION

Solving the first equation for ℓ and substituting the expression in the second equation gives an expression for h you can substitute in the third equation.

$$\ell w = 120$$

$$\ell = \frac{120}{w}$$

$$\ell h = 96$$

$$\left(\frac{120}{w}\right) h = 96$$

$$h = \frac{96w}{120}$$

$$= 0.8w$$

$$wh = 80$$
$$w(0.8w) = 80$$
$$0.8w^2 = 80$$
$$w^2 = 100$$
$$w = 10$$

$$h = 0.8w = 0.8(10) = 8$$

$$\ell = \frac{120}{w} = \frac{120}{10} = 12$$

The box is 12 units × 10 units × 8 units.

WARM-UP FOR PROBLEM 22

SOLUTION BY EXAMINING FACTORS

This approach will find only integer value solutions.

$\ell w = 120 = 2 \times 2 \times 2 \times 3 \times 5$
$\ell h = 96 = 2 \times 2 \times 2 \times 2 \times 2 \times 3$
$wh = 80 = 2 \times 2 \times 2 \times 2 \times 5$

Study the equations two at a time. The first two equations, $\ell w = 120$ and $\ell h = 96$, suggest the following.

- The common factor is $2 \times 2 \times 2 \times 3$, so ℓ must be some combination of some or all of these factors.
- Only ℓw has a factor of 5 (ℓh does not), so w must have a factor of 5.
- There are two 2's in ℓh that are not in ℓw, so h must have at least two 2's.

The second pair of equations, $\ell h = 96$ and $wh = 80$, gives these additional clues.

- The common factors are four 2's, so h has only factors of 2 and has at most four 2's.
- Only ℓh has a 3, so ℓ must have a factor of 3.
- There is one more 2 in ℓh than in wh, so ℓ must have at least one 2.
- Since wh has no 3's, neither w nor h has a factor of 3.

Next, study the first and third equations, $\ell w = 120$ and $wh = 80$.

- The common factor is $2 \times 2 \times 2 \times 5$, but one of the 2's in ℓw belongs to ℓ. Therefore, w is some combination obtained from $2 \times 2 \times 5$.

The possibilities can now be charted and their implications studied. The chart at the bottom of this page shows that the only possibilities that will work are for $\ell = 12$, $w = 10$, and $h = 8$.

Teaching Suggestions:

A few of my students don't want to solve this problem by solving equations. I encourage them to examine prime factors. (I do tell them that the dimensions are integer values.) Examining prime factors is an interesting solution method and you may wish to share it with your class if none of your students do.

For those students who wish to solve equations, mentioning the substitution method is enough to get them on their way.

POSSIBLE FACTORIZATIONS OF ℓ	IMPLICATIONS FROM FIRST EQUATION	SECOND EQUATION	THIRD EQUATION
$2 \times 2 \times 3$	$w = 5$	$h = 2 \times 2$	can't hold
$2 \times 2 \times 3$	$w = 2 \times 5$	$h = 2 \times 2 \times 2$	holds
2×3	$w = 2 \times 2 \times 5$	$h = 2 \times 2 \times 2 \times 2$	can't hold

FOUR SQUARE AND EIGHT

The midpoint of each side of a square is joined to its two opposite vertices. Each side of the square is four units long. What is the area of the octagonal region formed within the square?

Discussion for

FOUR SQUARE AND EIGHT

Answer: $2^{2}/_{3}$ square units

Clues:

1. The octagon is not a regular octagon since $MG = 1$ and $MF = (2/3)\sqrt{2}$.
2. Draw an accurate picture on graph paper.
3. Can you label all points of intersection? Try using coordinate geometry.
4. Separate the triangle into eight congruent triangles. Find the areas of the triangles.

Solution:

Figure $ABCD$ represents the square described in the problem. The midpoints of the sides are P, Q, R, and S. The octagonal region formed within the square has vertices E, F, G, H, I, J, K, and L. The center of both the square and the octagon is M.

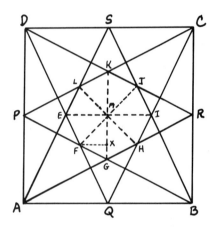

In this figure, all vertices of the octagon are connected to the center forming eight congruent triangles. The area of the octagonal region is eight times the area of one of these triangles. Using coordinate geometry, you can find the coordinates of the vertices of the triangles and hence the lengths of their bases and heights. For example,

equation for $\overleftrightarrow{PB}$: $m = \dfrac{2 - 0}{0 - 4}$

$= -\dfrac{1}{2}$

$b = 2$

$y = mx + b$

$= -\dfrac{1}{2}x + 2$

equation for $\overleftrightarrow{DQ}$: $m = \dfrac{4 - 0}{0 - 2}$

$= -2$

$b = 4$

$y = mx + b$

$= -2x + 4$

F is the intersection of $\overleftrightarrow{PB}$ and $\overleftrightarrow{DQ}$:

$y = -\dfrac{1}{2}x + 2$

$y = -2x + 4$

$-\dfrac{1}{2}x + 2 = -2x + 4$

$\dfrac{3}{2}x = 2$

$x = \dfrac{4}{3}$

$y = -2\left(\dfrac{4}{3}\right) + 4$

$y = \dfrac{4}{3}$

The coordinates of F are $(4/3, 4/3)$.

By using this coordinate geometry method (or by making a judicious choice of scale for your graph), you will find that the coordinates of $\triangle FGM$ are $(4/3, 4/3)$, $(2, 1)$, and $(2, 2)$, respectively. If you consider $\overline{MG}$ as the base of the triangle, then $\overline{FX}$ represents its height.

$MG = \sqrt{(2-2)^2 + (2-1)^2}$

$= \sqrt{1^2}$

$= 1$

X is $(2, 4/3)$.

$FX = \sqrt{\left(\dfrac{4}{3} - \dfrac{4}{3}\right)^2 + \left(2 - \dfrac{4}{3}\right)^2}$

$= 2 - \dfrac{4}{3}$

$= \dfrac{2}{3}$

Area $(\triangle FGM) = \dfrac{1}{2}(MG)(FX)$

$= \dfrac{1}{2}(1)\left(\dfrac{2}{3}\right)$

$= \dfrac{1}{3}$

Area (octagon) = $8 \times$ Area $(\triangle FGM)$

$= 8 \times \left(\dfrac{1}{3}\right)$

$= 2\dfrac{2}{3}$

The area of the octagon is $2^{2}/_{3}$ square units.

CONTINUED ON PAGE 154

REFLECT ON THIS

A square in the coordinate plane has vertices at (0, 0), (0, 4), (4, 4), and (4, 0). Three line segments are drawn starting at (0, 1) so that the first segment ends at the top, the second starts at that point and ends at the base, and the third starts at the base point and terminates at (4, 2). If you want the sum of the three segments to be the shortest possible measure, at what points do the segments meet the top and base of the square?

Discussion for

REFLECT ON THIS

Answer:

The segments meet the top of the square at (4/3, 4) and the base at (28/9, 0).

Clues:

1. The shortest distance between two points is a straight line.
2. On a billiard table the angle of inflection is congruent to the angle of deflection.
3. Suppose $(x_B, 4)$ is the point where the segments meet the top. What is the slope of the line joining (0, 1) and $(x_B, 4)$?
4. Suppose $(x_C, 0)$ is the base point. What is the slope of the line joining $(x_C, 0)$ and (4, 2)?
5. Draw a diagram on a piece of graph paper. Show several reflections of the square. (Recall your work on PROBLEM SET 9, particularly EXTENSION PROBLEM 9.)

Solution:

The diagram below shows the square and two reflections of it. The segments from A to B to C to D represent the segments described in the problem. The coordinates for B are $(x_B, 4)$ and for C are $(x_C, 0)$. The reflections as drawn show the path accurately because it preserves the angles of incidence and deflection.

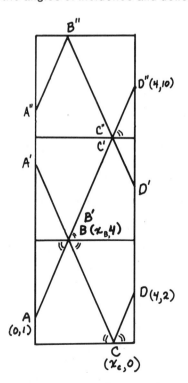

EXTENSION FOR PROBLEM 22

Notice that D'' is at (4, 10) and the slope of the segment from A to D'' is the same as the slopes of $\overline{AB}$ and $\overline{CD}$.

$$\text{slope } \overline{AD''} = \frac{10-1}{4-0}$$
$$= \frac{9}{4}$$

$$\text{slope } \overline{AB} = \frac{4-1}{x_B - 0}$$
$$= \frac{3}{x_B}$$

$$\text{slope } \overline{CD} = \frac{2-0}{4-x_C}$$
$$= \frac{2}{4-x_C}$$

$$\frac{3}{x_B} = \frac{9}{4} \qquad \frac{2}{4-x_C} = \frac{9}{4}$$

$$9x_B = 12 \qquad 36 - 9x_C = 8$$

$$x_B = \frac{12}{9} \qquad 9x_C = 28$$

$$= \frac{4}{3} \qquad x_C = \frac{28}{9}$$

The segments meet the top of the square at (4/3, 4) and the base of the square at (28/9, 0).

Teaching Suggestions:

Drawing an accurate picture is very helpful.

This problem reinforces what students have learned about coordinate geometry in Algebra One.

DOUBLE OR NOTHING

To boost sales, the president of Slirpy Soups decided to change the size of its soup can by doubling one dimension and halving the other. Not knowing which dimension to double and which to halve, the production people created cans of both types; they made cans with doubled heights and halved radii and they made cans with halved heights and doubled radii.

Both cans sold for the same price—the same price as the original can of Slirpy Soup. One of the new cans sold very well, but the other did not. Why?

Discussion for
DOUBLE OR NOTHING

Answer:

The cans with the doubled heights and halved radii did not sell very well because their volumes were half those of the original can; they held *half* as much Slirpy Soup. The cans with halved heights and doubled radii sold very well; their volumes were *twice* the volume of the original cans.

Clues:

1. Find volumes. (The formula for the volume of a cylinder is $V = \pi r^2 h$ where r is the radius and h is the height.)
2. Suppose the original can had a height of 4 units and a radius of 2 units.
3. How does the square of a fraction compare to the fraction?

Solution:

Let h stand for the height of the original can and let r stand for its radius. Then its volume is $\pi r^2 h$.

Now find the volumes for the two new cans and compare.

HEIGHT DOUBLED, RADIUS HALVED

$$\text{Volume} = \pi \,(\text{radius})^2 \,(\text{height})$$
$$= \pi \left(\frac{r}{2}\right)^2 (2h)$$
$$= \frac{1}{2} \pi r^2 h$$
$$= \frac{1}{2} \,(\text{volume original can})$$

HEIGHT HALVED, RADIUS DOUBLED

$$\text{Volume} = \pi \,(\text{radius})^2 \,(\text{height})$$
$$= \pi \,(2r)^2 \left(\frac{h}{2}\right)$$
$$= 2\pi r^2 h$$
$$= 2 \,(\text{volume original can})$$

Teaching Suggestions:

Some students have difficulty generalizing; I encourage them to start out with specific values for height and radius, trying several examples in order to get a feeling for the problem.

Given the conditions of this problem, you may wish to have students answer the following questions.

- Should the price of the new cans be the same as the original? (No.)
- To maintain the original volume, what conditions must the radius meet if the height is doubled? (It must be $\sqrt{2}/2$ times the original radius.)
- To maintain the original volume, what conditions must the height meet if the radius is doubled? (It must be 1/4 the original height.)
- Suppose the price of the original can was $0.49. What price should a new can have if both its height and radius are doubled and you want to maintain your profit margin? ($3.92.)

WARM-UP FOR PROBLEM 23

MULTI-PI

The area and volume of a given sphere are four-digit integer multiples of pi. What is the radius of the sphere?

Discussion for

MULTI-PI

Answer: 18 units

Clues:

1. The formula for the area of a sphere is $A = 4\pi r^2$.
2. The formula for the volume of a sphere is $V = (4/3)\pi r^3$.
3. Find the values of r, the radius of the sphere, that satisfy the following inequality.
   $1000\pi \leq 4\pi r^2 < 10{,}000\pi$
4. What conditions must r meet to ensure that $(4/3)\pi r^3$ is an integer?
5. The value of the radius must be an integer.

Solution:

The least four-digit integer is 1000 and the greatest is one less than 10,000. The conditions of the problem can be stated as a series of inequalities that can be simplified (letting r stand for the radius of the sphere).

AREA INEQUALITIES

$$1000\pi \leq 4\pi r^2 < 10{,}000\pi$$
$$1000 \leq 4r^2 < 10{,}000$$
$$250 \leq r^2 < 2500$$
$$\sqrt{250} \leq r < \sqrt{2500}$$
$$5\sqrt{10} \leq r < 50$$

VOLUME INEQUALITIES

$$1000\pi \leq \frac{4}{3}\pi r^3 < 10{,}000\pi$$
$$\frac{3}{4}(1000) \leq r^3 < \frac{3}{4}(10{,}000)$$
$$750 \leq r^3 < 7500$$
$$5\sqrt[3]{6} \leq r < 5\sqrt[3]{60}$$

The values for area and volume must be *integer* multiples of π, so $4r^2$ and $(4/3)r^3$ must be integers. These conditions imply that r must be an integer and, furthermore, r must be a multiple of 3.

Since $5\sqrt{10} \approx 15.8$ and $5\sqrt[3]{6} \approx 9.09$, r must be greater than or equal to 16. Since $5\sqrt[3]{60} \approx 19.58$, r must be less than 20. And the only integer value from 16 to 20 that is a multiple of 3 is 18, so r is 18 (giving an area of 1296π units^2 and a volume of 7776π units^3).

Teaching Suggestions:

Students could solve this problem by trial and error, but the inequality method is much more efficient. Helping students set up one of the conditions (for example, the least four-digit integer is 1000, so the area is greater than or equal to 1000π) is enough to get them started. You may also wish to point out that approximations for π (such as 3.14 or 22/7) are not needed here, since π divides out.

GEOMETRY PROBLEM 23

FILL IT TO THE RIM

A coffee pot with a circular bottom tapers uniformly to a circular top having radius half that of the base. A mark halfway up (by height) says *2 cups*. If the pot could be filled completely to the rim, how much coffee would it hold?

Discussion for
FILL IT TO THE RIM

Answer: 3-1/37 cups

Clues:

1. The pot is the bottom part of what figure?
2. Find the volume of the entire figure (of the cone).
3. The ratio of the volumes of two similar cones is directly proportional to the cube of the ratio of the corresponding heights of the cones.
4. Suppose h represents the height of the cone. Find an expression for the distance from the tip of the cone to the inside of the cone at the 2 cup mark.

Solution:

The triangle in the following diagram shows the cross section of a cone with height h and radius r. Figure ACEG represents a cross section of the coffee pot, so $CJ = (1/2)r$. $\overline{BF}$ shows where the 2 cup mark cuts across the pot.

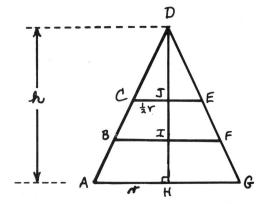

The cones with cross sections $\triangle CDE$, $\triangle BDF$, and $\triangle ADG$ are similar. Therefore, their heights and radii are proportional. Using this fact and information from the problem statement, you can make the following deductions.

Since the radius of the top part of the pot equals one-half the radius of the pot, the pot is the frustum of a cone that has a height twice that of the pot.

The volumes of two similar cones are proportional to the cubes of the heights of the two cones. Therefore, if V represents the volume of the entire cone and $V - 2$ represents the volume of the cone whose base is at the 2 cup mark, then you can write the following proportion. (Let 1 unit represent the height of the large cone; then the smaller cone has height ¾ unit.)

$$\frac{V}{V-2} = \frac{1^3}{\left(\frac{3}{4}\right)^3}$$

$$\left(\frac{27}{64}\right)V = V - 2$$

$$V = \frac{128}{37}$$

You can find the volume of the cut-off portion of the cone in the same way. (Let V_c stand for the volume.)

$$\frac{V}{V_c} = \frac{1^3}{\left(\frac{1}{2}\right)^3}$$

$$V_c = \frac{1}{8}(V)$$

$$= \frac{1}{8}\left(\frac{128}{37}\right)$$

$$= \frac{16}{37}$$

Therefore, the volume of the pot is $(128/37) - (16/37)$, or 3-1/37.

Teaching Suggestions:

The key to solving this problem is in understanding the relationships between similar figures (in particular in understanding that the volume ratio is proportional to the *cube* of the ratio of sides, or radii, or heights). You may wish to warm students up to the idea by first working with some simpler problems. For example, if the edge of one cube is half the edge of another and their volumes differ by 56 units³, what are the volumes of the two cubes?

If x stands for the volume of the large cube, then the volume of the smaller cube is $x - 56$ and you can set up and solve the following proportion.

$$\frac{x}{x-56} = \frac{2^3}{1^3}$$

$$x = 8x - 448$$

$$7x = 448$$

$$x = 64$$

$$x - 56 = 8$$

Students will have a much easier time with the given problem if they first draw a diagram and find the volumes of the three cones.

EXTENSION FOR PROBLEM 23

TWO FOR ONE

A cube made from 1-cm cubes measures 6 cm along each edge. The cube is rearranged to form two rectangular prisms. The base of one prism is 8 cm × 5 cm. The base of the other prism is 6 cm × 4 cm. What is the height of each prism?

Discussion for

TWO FOR ONE

Answer:

The prism whose base is 8 cm × 5 cm has a height of 3 cm. The prism whose base is 6 cm × 4 cm has a height of 4 cm.

Clues:

1. Can the heights have fractional values?
2. What is the combined volume of the two prisms?
3. The heights of the two prisms are not the same.
4. Try guessing and checking.

Solution:

The original cube is 6 cm × 6 cm × 6 cm for a total of 216 one-centimeter cubes. If x stands for the height of the prism with the 8 cm × 5 cm base, then its volume is $40x$ cm^3. Similarly, if y stands for the height of the prism having a 6 cm × 4 cm base, then its volume is $24y$ cm^3. The two volumes together must add to 216.

$$40x + 24y = 216$$
$$8(5x + 3y) = 8(27)$$
$$5x + 3y = 27$$
$$y = \frac{1}{3}(27 - 5x)$$

You are dealing with one-centimeter cubes, so the values of x and y must be integers. (The equation is a Diophantine equation.)

A guess-and-check method gives you the solution very quickly, as the following chart shows.

x	$y = \frac{1}{3}(27 - 5x)$
1	$\frac{22}{3}$
2	$\frac{17}{3}$
3	4
4	$\frac{7}{3}$
5	$\frac{2}{3}$

The only possible integer values for x and y are 3 and 4, respectively.
The only possible integer values for x and y are 3 and 4, respectively.

Teaching Suggestions:

Most students will solve this equation by trial and error. However, they will appreciate seeing an analysis of $(1/3)(27 - 5x)$. That is, that $27 - 5x$ must be divisible by 3, so the only possible value for x must have a factor of 3, implying that $x = 3$ (since other multiples of 3 yield a negative value).

WARM-UP FOR PROBLEM 24

REVOLUTIONARY FIGURES

By revolving a 6 unit × 8 unit × 10 unit triangle about one of its sides, a solid figure can be produced. Each side creates a different figure. Which of the three possible figures has the greatest volume? the least volume?

Discussion for
REVOLUTIONARY FIGURES

Answer:

The figure formed by revolving about the 6 unit side has the greatest volume, 128π units^3. The figure formed by revolving about the 10 unit side (the hypotenuse) has the least volume, 76.8π units^3. Revolving about the third side, the 8 unit side, produces a figure with a volume of 96π units^3.

Clues:

1. What kind of triangle is involved?
2. The volume of a cone is given by $V = (1/3)Bh$ where B stands for the area of the circular base and h stands for the height.
3. What theorems have you learned about the altitude of a right triangle drawn from the right angle to the hypotenuse?
4. Two of the figures produced are right circular cones (whose axes of revolution are legs of the given right triangle).
5. The third figure resembles two right circular cones glued together at their bases. (The axis of revolution is the hypotenuse of the given right triangle.)

Solution:

The measures of the sides of the given triangle are a Pythagorean triple, so the triangle is a right triangle with a 10 unit hypotenuse.

$$6^2 + 8^2 = 36 + 64$$
$$= 100$$
$$= 10^2$$

When the triangle is revolved about either leg, a right circular cone is produced. The axis of revolution is the height.

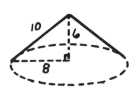

$$V = \frac{1}{3}\pi r^2 h \qquad V = \frac{1}{3}\pi r^2 h$$
$$= \frac{\pi}{3}(8^2)(6) \qquad = \frac{\pi}{3}(6^2)(8)$$
$$= 128 \qquad = 96$$

When the 10 unit hypotenuse is the axis of revolution, a double right circular cone is produced. Let a stand for its radius and x and $10 - x$ stand for the heights of the two cones as shown in the diagram.

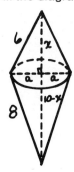

In a right triangle, when the altitude is drawn from the right angle to the hypotenuse, then the triangles formed are similar to the original right triangle and the altitude is the geometric mean between the measures of the two segments formed along the hypotenuse. These facts allow you to write the following two proportions.

$$\frac{x}{6} = \frac{6}{10} \qquad \frac{x}{a} = \frac{a}{10-x}$$
$$x = 6\left(\frac{6}{10}\right) \qquad a^2 = x(10-x)$$
$$= 3.6 \qquad = (3.6)(10 - 3.6)$$
$$\qquad\qquad = 23.04$$

The volume of the figure is obtained by adding the volumes of the two cones.

$$V = \frac{1}{3}(\pi a^2)(x) + \frac{1}{3}(\pi a^2)(10 - x)$$
$$= \frac{1}{3}\pi a^2 (x + 10 - x)$$
$$= \frac{10}{3}\pi a^2$$
$$= \frac{10}{3}\pi (23.04)$$
$$= 76.8\pi$$

Teaching Suggestions:

The figures created by revolving the triangle are solid figures. Quickly revolving a cardboard model of the triangle about each side helps make that point.

This problem was purposely constructed with a right triangle; however, any triangle could have been used. You find the altitude by writing two different equations about the altitude and solving them simultaneously. If you present this problem using obtuse triangles, you will give your students valuable practice in figure sketching.

With the 6 × 8 × 10 triangle, the hypotenuse created the figure with the least volume. You may wish to investigate the general question, "Does the hypotenuse always create the figure with the least volume?" The answer is yes.

GEOMETRY PROBLEM 24

DRILL BIT

If a hole 6 cm in length from opening to opening is drilled through the center of a solid sphere that is more than 6 cm in diameter, what is the volume of the remaining portion of the sphere?

Discussion for

DRILL BIT

Answer: 36π cm³

Clues:

1. The height of the cylindrical portion is 6 cm.
2. The formula for the volume of a sphere is $V = (4/3)\pi r^3$ where r is the radius of the sphere.
3. The formula for the cap of a sphere is $V = (1/3)\pi h^2(3r - h)$ where r is the radius of the sphere and h is the height of the cap. (Students must have access to this information; they won't be able to derive it.)
4. Express a value for the radius of the cylinder in terms of the radius of the sphere. (Use the Pythagorean Theorem.)

Solution:

In the following cross-sectional diagram of the remaining portion of the drilled sphere, R stands for the radius of the cylinder, r stands for the radius of the sphere. Since the cylinder is 6 cm tall, the height of the cap is $r - 3$.

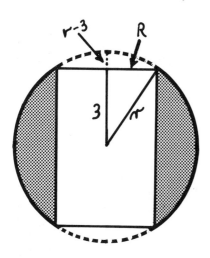

By using the Pythagorean Theorem, you can find a value for R^2 in terms of r and, hence, an expression for the volume of the cylinder in terms of r.

$$r^2 = R^2 + 3^2$$
$$R^2 = r^2 - 9$$
$$R = \sqrt{r^2 - 9}$$

Volume (cylinder) $= \pi$ (radius)² (height)
$$= \pi (r^2 - 9)(6)$$

Since the height of each cap is $r - 3$, their volumes are as follows. (The formula is given in most standard references. It is derived using calculus.)

Volume (cap) $= \dfrac{1}{3}\pi$ (height)² [3 (sphere radius) $-$ height]

$$= \dfrac{1}{3}\pi (r - 3)^2 [3r - (r - 3)]$$

$$= \dfrac{1}{3}\pi (r - 3)^2 (2r + 3)$$

$$= \dfrac{\pi}{3}(2r^3 - 9r^2 + 27)$$

The desired volume can be calculated as follows.

Remaining Volume = Volume Sphere $-$ Volume Cylinder $- 2$(Volume Cap)

$$= \left[\dfrac{4}{3}\pi r^3\right] - [6\pi(r^2 - 9)]$$
$$\quad - 2\left[\dfrac{\pi}{3}(2r^3 - 9r^2 + 27)\right]$$

$$= \dfrac{4}{3}\pi r^3 - 6\pi r^2 + 54\pi$$

$$\quad - \dfrac{4}{3}\pi r^3 + 6\pi r^2 - 18\pi$$

$$= 36\pi$$

Teaching Suggestions:

As usual, I find that some students need to review squaring binomials in order to expand the expression for the volume of the caps. The algebra involved in this expression alone gives the much-needed reinforcement for the forgotten skills.

The final answer is independent of the radius of the sphere (and of the cylinder radius). Going through the solution step by step usually makes the point.

EXTENSION FOR PROBLEM 24

GOT YOU CORNERED

Suppose point P is located in the interior of a rectangle so that the distance from P to a corner of the rectangle is 5 m, the distance to the opposite corner is 14 m, and to a third corner is 10 m. What is the distance from P to the fourth corner?

Discussion for

GOT YOU CORNERED

Answer: 11 m

Clues:

1. Draw an accurate sketch. (*P* is not necessarily on the diagonal of the square.)
2. **Separate the rectangle into four retangular parts using the given point as a common vertex.**
3. Express each distance from *P* to a vertex in terms of parts of the sides of the rectangle.
4. Write four different expressions using the Pythagorean Theorem.
5. Combine the expressions from Clue 3 to make one equation having one unknown.

Solution:

In the following diagram, *P* is the given point, *BP*, *DP*, and *AP* are the given distances, and *CP* is the distance to be found. Lines are drawn through *P* perpendicular to the sides of the rectangle. They form four different rectangles and each segment ($\overline{BP}, \overline{DP}, \overline{AP}, \overline{CP}$) is a diagonal.

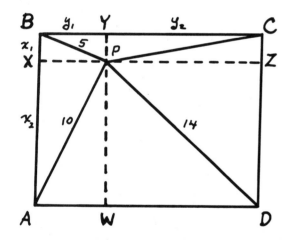

By using the Pythagorean Theorem, you can write the following four equations.

$$x_1^2 + y_1^2 = BP^2 \qquad x_1^2 + y_2^2 = CP^2$$
$$x_2^2 + y_2^2 = DP^2 \qquad x_2^2 + y_1^2 = AP^2$$

Adding pairs of equations to obtain $x_1^2 + x_2^2 + y_1^2 + y_2^2$ allows you to make an equation about the diagonals.

$$x_1^2 + x_2^2 + y_1^2 + y_2^2 = BP^2 + DP^2$$
$$x_1^2 + x_2^2 + y_1^2 + y_2^2 = CP^2 + AP^2$$
$$BP^2 + DP^2 = CP^2 + AP^2$$

In the case of this problem, $AP = 10$, $BP = 5$, and $DP = 14$.

$$5^2 + 14^2 = CP^2 + 10^2$$
$$CP^2 = 25 + 196 - 100$$
$$= 121$$
$$CP = 11$$

Teaching Suggestions:

The solution given here presents a general case; not until the end are the actual values given in the problem statement used. The solution, in fact, shows that for *any* point interior to a rectangle, the sum of the squares of the distances from the point to a set of opposite corners equals the sum of the squares of the distances to the other set of corners. As an extension to the given problem, you may wish to have students prove this general case.

Be very careful in generating other specific values for this problem. An arbitrary choice of values could result in a nonexistent answer.

SUPER MATCH

What is the least number of matchsticks needed for the following construction?

- Form a square out of matchsticks.
- Construct one right triangle on each side of the square.
- Make sure each triangle has different dimensions.
- Do not break any matchsticks to make the figures.
- The hypotenuse of a triangle cannot be a side of the square!

Discussion for

SUPER MATCH

Answer: 198 matchsticks

Clues:

1. One side of each triangle must be the same length.
2. If s is the number of matchsticks along the side of the square, h is the number of matchsticks along the hypotenuse of a triangle, and ℓ is the number of matchsticks along the other leg of the triangle, then $s^2 = h^2 - \ell^2$.
3. Find four different sets of Pythagorean triples that satisfy the equation in Clue 2, each with the same value for s.
4. The area of the square is an even number of square units.
5. The number of matchsticks along the side of the square equals twice the first perfect number.
6. For $x < y$, the expressions $y^2 - x^2$, $2xy$, and $x^2 + y^2$ all represent numbers that can be sides of a right triangle.

Solution:

If s is the number of matchsticks along the side of the square, h is the number of matchsticks along the hypotenuse of a triangle, and ℓ is the number of matchsticks along the other leg of the triangle, then $s^2 = h^2 - \ell^2$. Since there are four *different* triangles, the problem is first to find four different sets of Pythagorean triples, with the same value for s in each.

The following list contains all the Pythagorean triples in which no number is greater than 50.

3, 4, 5	12, 16, 20	21, 28, 35
5, 12, 13	12, 35, 37	24, 32, 40
6, 8, 10	14, 48, 50	27, 36, 45
7, 24, 25	15, 20, 25	30, 40, 50
8, 15, 17	15, 36, 39	
9, 12, 15	16, 30, 34	
9, 40, 41	18, 24, 30	
10, 24, 26	20, 21, 29	

From this list there are three sets of four triples having a common number. Of these sets, only two meet the conditions of the problem (the common number cannot be the hypotenuse of the triangle).

SET 1	SET 2	SET 3
7, 24, 25	5, 12, 13	8, 15, 17
10, 24, 26	9, 12, 15	9, 12, 15
18, 24, 30	12, 16, 20	15, 20, 25
24, 32, 40	12, 35, 37	15, 36, 39

GEOMETRY PROBLEM 25

The number of matchsticks in the entire figure can be obtained by totaling the numbers in the sets.

MATCHSTICKS USING SET 1	MATCHSTICKS USING SET 2
284	198

Any other sets of Pythagorean triples, with numbers greater than 50, would clearly give a larger total than those for Sets 1 and 2. (The starting numbers are greater and addition preserves inequality.) Therefore, the least number of matchsticks needed is 198, to form a 12 × 12 square with 12 × 5 × 13, 12 × 9 × 15, 12 × 16 × 20, and 12 × 35 × 37 triangles on each side, respectively.

Teaching Suggestions:

If you take Clue 4 as a given, you can go a long way toward solving this problem by examining factors. The right-hand side of the equation, $h^2 - \ell^2$, is the difference of two squares and can be factored to $(h - \ell)(h + \ell)$. Since h and ℓ are integers, the factors must be either both even or both odd. But the Clue 4 condition implies that the factors must both be even.

h	ℓ	$h - \ell$	$h + \ell$	s^2
even	even	even	even	even
even	odd	odd	odd	odd
odd	even	odd	odd	odd
odd	odd	even	even	even

And, if both $(h - \ell)$ and $(h + \ell)$ are even, then s^2 must have a factor of 4, so s must be an even number. Then you can complete the problem by searching for pairs of factors for $(h - \ell)$ and $(h + \ell)$ that give appropriate values for s^2. You will quickly come upon 12.

Once you have completed an analysis for an even area, you can carry out a similar analysis for an odd area. The first set of triangles you will come upon generates triangles with dimensions 15 × 112 × 114, 15 × 36 × 39, 15 × 20 × 25, and 15 × 8 × 17, giving a total of 430 (which is much greater than necessary).

Factoring in this way is a very useful method for a number of different problems. Having students start with small values of s and asking them to find appropriate values for h and ℓ gives practice both in factoring integers and in solving simultaneous equations.

With the exception of $s = 1$, it is always possible to find three rational numbers that can make up a right triangle (where s is an integer). For example, if $s = 2$, then $s^2 = 4 = (1)(4)$, so $h - \ell = 1$ and $h + \ell = 4$. Solving simultaneously results in values of 2.5 for h and 1.5 for ℓ.

The use of home computers turns this problem into a sorting problem once a list of possible Pythagorean Triples has been produced.

WHATEVER'S RIGHT

One leg of a right triangle is 48 units long and the other two sides have integral lengths. Find all the possible lengths of the two sides.

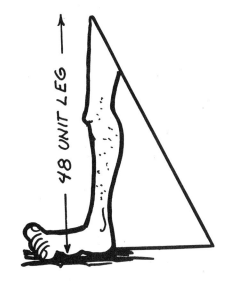

Discussion for
WHATEVER'S RIGHT

Answer:

The hypotenuse and leg are given by the following pairs.

577, 575	80, 64
290, 286	73, 55
195, 189	60, 36
148, 140	52, 20
102, 90	50, 14

Clues:

1. There are ten possible answers.
2. List all the two-number factorizations that give a product of 48^2.
3. If $x^2 + y^2 = h^2$, then $y^2 = h^2 - x^2$.
4. What is the factorization of $h^2 - x^2$?
5. If h and x are positive numbers with $h > x$, then $h + x$ is greater than $h - x$.
6. Both $(h + x)$ and $(h - x)$ must represent even numbers.
7. Solve simultaneous equations.

Solution:

The diagram below shows the triangle; h represents the length of the hypotenuse and x the length of the other leg.

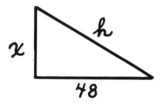

By the Pythagorean Theorem, the sides are related as follows.

$$48^2 = h^2 - x^2$$
$$= (h - x)(h + x)$$

The first step in solving the problem becomes finding the factorizations of 48^2. (The prime factorization is $2^8 \cdot 3^2$, so you can find them all by checking all the terms of the product $(1 + 2 + 2^2 + 2^3 + \cdots + 2^8)(1 + 3 + 3^2)$.) The factorizations are:

1 × 2304	12 × 192
2 × 1152	16 × 144
3 × 768	18 × 128
4 × 576	24 × 96
6 × 384	32 × 72
8 × 288	36 × 64
9 × 256	48 × 48

Both factors must be even because, as the chart shows, only when both $(h - x)$ and $(h + x)$ are even is $h^2 - x^2$ even.

h	x	$h - x$	$h + x$	$h^2 - x^2$
even	even	even	even	even
even	odd	odd	odd	odd
odd	even	odd	odd	odd
odd	odd	even	even	even

Therefore, only 10 of the factorizations are possible.

By solving the pairs of equations formed, you will find the possible lengths of the two sides. For example, when $h - x = 12$ and $h + x = 192$, $h = 102$ and $x = 90$.

Teaching Suggestions:

Even though this problem is an extension of the ideas in GEOMETRY PROBLEM 25, you may wish to present this one first. In both problems, the key to the solution is finding factors for $(h - x)(h + x)$ and then using these factors to create pairs of equations to be solved simultaneously.

The mathematics of this problem is quite interesting but a little tedious. Don't despair if your students are not willing to stick with it long enough to get all the answers. You might even suggest that they find one, two, or three possible solutions instead.

By using similar triangles and basic Pythagorean triples (like 3-4-5) many solutions can be found very quickly. (One solution is the triple 36-48-60 which is derived from 3-4-5.)

EXTENSION FOR PROBLEM 25

STAIR STEPS continued

Teaching Suggestions:

Once students have begun making a table that relates the two values T and n, most will be able to continue without help, until they can "bracket" the number 350. Be sure they answer the question asked by the problem.

I think that the technique of finite differences is particularly useful for a number of sequence problems. Don't be alarmed if most of your students do not follow an explanation of the technique the first time they are exposed to it. Understanding will come with practice. (I usually introduce finite differences early in the year with my Algebra One classes and expand upon the technique when we begin to solve simultaneous equations.) Whenever I introduce finite differences, I usually provide copies of charts like those in the back of this book, allowing students to use them on this problem and on subsequent similar problems.

RICOCHET continued

Teaching Suggestions:

This problem is *not* about billiards; it only involves one ball. Students need to understand that the problem does *not* involve hitting one ball into another.

You will find that talking about the angle of incidence and the angle of reflection is usually enough discussion to get students started on their way to a solution.

CUBED SALAD continued

Students might also wish to experiment with a regular tetrahedron and try to find the different shapes, although the possibilities are more limited than for a cube.

When working with the cube, I like to stump my students with the following question.

> Is it possible to cut a hole through a cube in such a way that a larger cube can be passed throught the hole?

Most students answer *no* without thinking very hard, because they are relating the problem to spheres. Consider the following figure in which the edges of the cube are all one unit in length.

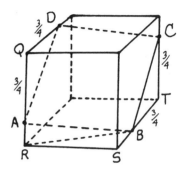

Two points are located 3/4 unit from vertex Q. Also, two points are located 3/4 unit from vertex T. Connecting these points forms a square that has sides of length $(3/4)\sqrt{2}$, or about 1.06 units! ABCD is a rectangle since opposite sides are parallel and since diagonals $\overline{AC}$ and $\overline{BD}$ are congruent by the symmetry of the cube and, thus, bisect each other at the center of the cube. Since triangle ADQ is an isosceles right triangle with legs 3/4 units long, $AD = (3/4)\sqrt{2}$. The same result holds for BC. The Pythagorean Theorem implies that $AB^2 = AR^2 + BR^2$ and that $BR^2 = RS^2 + BS^2$, so $AB^2 = AR^2 + RS^2 + BS^2 = (1/4)^2 + 1 + (1/4)^2 = 18/16$, making AB equal to $(3/4)\sqrt{2}$. Hence, ABCD is a rectangle with sides of equal length—a square.

TWO'S COMPANY continued

TWO CONSECUTIVE VERTICES (8 segments)

$\overline{V_1-V_2}, \overline{V_2-V_3}, \overline{V_3-V_4}, \overline{V_4-V_5}, \overline{V_5-V_6}, \overline{V_6-V_7}, \overline{V_7-V_8}, \overline{V_8-V_1}$

$5 + 4 + 3 + 2 + 1 = 15 \times 8$ cases = 120 triangles

Examples:
$\overline{V_1-V_2}$ →

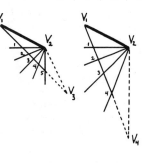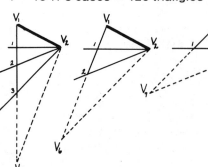

SKIP ONE VERTEX (8 segments)

$\overline{V_1-V_3}, \overline{V_2-V_4}, \overline{V_3-V_5}, \overline{V_4-V_6}, \overline{V_5-V_7}, \overline{V_6-V_8}, \overline{V_7-V_1}, \overline{V_8-V_2}$

$4 + 3 + 2 + 1 = 10 \times 8$ cases = 80 triangles

Examples:
$\overline{V_1-V_3}$ →

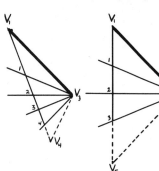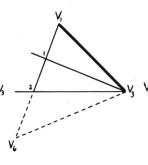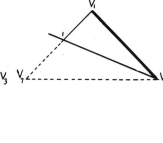

SKIP TWO VERTICES (8 segments)

$\overline{V_1-V_4}, \overline{V_2-V_5}, \overline{V_3-V_6}, \overline{V_4-V_7}, \overline{V_5-V_8}, \overline{V_6-V_1}, \overline{V_7-V_2}, \overline{V_8-V_3}$

$1 + 3 + 2 + 1 = 7 \times 8$ cases = 56 triangles

Examples:
$\overline{V_1-V_4}$ →

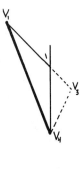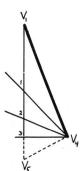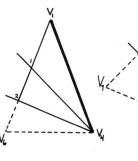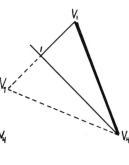

OPPOSITE VERTICES (4 segments)

$\overline{V_1-V_5}, \overline{V_2-V_6}, \overline{V_3-V_7}, \overline{V_4-V_8}$

$1 + 2 + 2 + 1 = 6 \times 4$ cases = 24 triangles

Examples:
$\overline{V_1-V_5}$ →

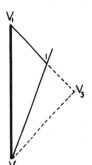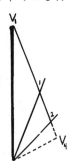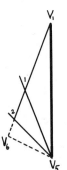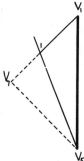

Sum total: 120 + 80 + 56 + 24 = 280 different triangles

POPSICLE STICK CONSTRUCTIONS continued

A LINE THROUGH A POINT AND PERPENDICULAR TO A GIVEN LINE

If the point is on the line, then mark points on the line, A and B, on either side of the given point and the straightedge's width away. Then construct two sets of three parallel lines through the three points. The lines through A and B will form a rhombus with P at its center. The two vertices not on the line can be connected to form a diagonal of the rhombus passing through P. This diagonal is perpendicular to diagonal $\overline{AB}$ and, hence, the entire line.

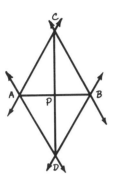

If the point is not on the line, first construct any line perpendicular to the given line (previous paragraph). Then construct a line through the given point parallel to the perpendicular you just constructed.

Teaching Suggestions:

A short discussion on projections is useful. You might start by drawing a line that is divided into two parts in a given ratio on the chalkboard or overhead. Pick any point that is not on the line and connect the three endpoints on your first line to that point. Then draw another line that is parallel to the original line and intersects the three connectors. Ask students to estimate the ratio of the segments formed on the new parallel.

Two of the simplest constructions that are typically introduced with a compass and straightedge were purposely omitted from this problem—copying a line segment and copying an angle. I do not believe that the work involved with this straightedge-only approach is worth the effort.

For students who become really involved with these constructions, looking into the Mascheroni constructions is a worthwhile extra credit project.

CANNONBALL RUN continued

The result is a layer having 6 rows of 9 cannonballs each alternating with 5 rows of 10 cannonballs each, for a total of 104 balls.

By alternating the layers of 105 balls with layers of 104 balls, Rufus can fit a total of 1254 cannonballs ($6 \cdot 105 + 6 \cdot 104$).

Teaching Suggestions:

The problem in its entirety is too difficult for most of my students. I usually accept any of the three answers derived in the solution, providing the students show their work. If students have been exposed to WARM-UP PROBLEM 10, the sphere-packing problem, the idea of a triangular-array packing will come easily.

The difficult part of this problem is in determining how many layers of cannonballs will fit into the box if each successive layer occupies the gap between the balls. The problem can be solved if it is possible to find the vertical interval between the axes of successive layers. The distance is approximately the height of the tetrahedron formed by joining the centers of four of the spheres, three in one layer and one in the next layer, tangent to the other three. (This problem does not involve a true tetrahedron, but the actual difference is negligible.)

Finding the height of the tetrahedron is in itself a good problem. If e represents the edge of a tetrahedron, then $(e/3)\sqrt{6}$ represents its height. Since the radius of a cannonball is 3 in., the edge of the tetrahedron is 6 in. and the height is $2\sqrt{6}$ in., or approximately 4.90 in. Twelve layers will reach a height of ($11 \cdot 4.90 + 6$) in., or about 59.9 in.

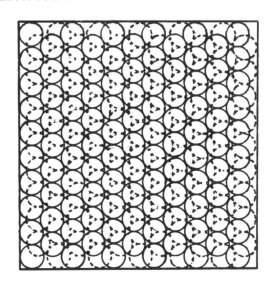

MAKING ARRANGEMENTS continued

Teaching Suggestions:

Using WARM-UP PROBLEM 16 is usually enough to get students started on this problem. Especially if you do not use the warm-up problem, you may want to suggest that students make a model of the problem and start out with some trial-and-error arrangements so they get a feel for the problem. Be sure students understand that, once they have packed spheres in a rectangular fashion, their borders are set; they cannot go outside those borders with their rearrangements.

If you use an overhead projector, you might wish to keep an ample supply of pennies handy. They are very helpful for showing various packing arrangements.

FOUR SQUARE AND EIGHT continued

Teaching Suggestions:

Drawing an accurate picture on graph paper is a helpful strategy with many kinds of problems that involve a geometric diagram. Students see more clearly how various parts of a diagram are related to one another and it can suggest the way to proceed.

If students are lucky enough to choose a multiple of three squares for each unit, they will easily discover the points of intersection that create the octagon. Students who choose another scale may need a few clues to get them going. They may realize that knowing the length of $\overline{FX}$ or the length of $\overline{FH}$ or a related length will allow them to solve the problem, but they may not know how to find those lengths. Rather than simply telling students to use coordinate geometry, you may find that they will benefit from an in-class discussion on different ways to get started.

Once students have found the key information for this problem, there are a number of different ways they can complete the problem. Encourage them to find more than one way. They will devise some ingenious plans.

FINITE DIFFERENCES
Linear Equations

GENERAL TABLE

x	$ax + b$	first difference
1	$a + b$	
		a
2	$2a + b$	
		a
3	$3a + b$	
		a
4	$4a + b$	
		a
5	$5a + b$	
.		
.		
.		
n	$na + b$	

EXAMPLE TABLE

n	f_n	first difference
1	5	
		2
2	7	
		2
3	9	
		2
4	11	
		2
5	13	
.		
.		
.		
n	$?n + ?$	

When a sequence gives a common difference after one subtraction, its nth term is linear. In this example, the sequence 5, 7, 9, 11, 13, . . . has a common difference of 2.

To find the nth term, solve for a and b. Always the value of a is the first difference.

$a = 2$

Use this value to help you find b. The first term of the sequence equals $a + b$.

$a + b = 5$
$(2) + b = 5$
$b = 3$

The nth term is $an + b$ which, in this case, is $2n + 3$.

FINITE DIFFERENCES
Second-Degree Equations

x	$ax^2 + bx + c$	first difference	second difference
1	$a + b + c$		
		$3a + b$	
2	$4a + 2b + c$		$2a$
		$5a + b$	
3	$9a + 3b + c$		$2a$
		$7a + b$	
4	$16a + 4b + c$		$2a$
		$9a + b$	
5	$25a + 5b + c$		
.			
.			
.			
n	$an^2 + bn + c$		

n	f_n	first difference	second difference
1	4		
		5	
2	9		2
		7	
3	16		2
		9	
4	25		2
		11	
5	36		
.			
.			
.			
n	$?n^2 + ?n + ?$		

When a sequence gives a common difference after two subtractions, its nth term is quadratic (is a second-degree expression). In this example, the sequence 4, 9, 16, 25, 36, . . . has a common difference of 2 after two subtractions.

To find the nth term, solve for a, b, and c. The value of a is always half the second difference.

$2a = 2$
$a = 1$

Look at the first difference to find a value for b.

$3a + b = 5$
$3(1) + b = 5$
$b = 2$

Look at the first term to find a value for c.

$a + b + c = 4$
$(1) + (2) + c = 4$
$c = 1$

The nth term is $an^2 + bn + c$ which, in this case, is $n^2 + 2n + 1$.

FINITE DIFFERENCES
Third-Degree Equations

GENERAL TABLE

x	$ax^3 + bx^2 + cx + d$	first difference	second difference	third difference
1	$a + b + c + d$			
		$7a + 3b + c$		
2	$8a + 4b + 2c + d$		$12a + 2b$	
		$19a + 5b + c$		$6a$
3	$27a + 9b + 4c + d$		$18a + 2b$	
		$37a + 7b + c$		$6a$
4	$64a + 16b + 4c + d$		$24a + 2b$	
		$61a + 9b + c$		
5	$125a + 25b + 5c + d$			
.				
.				
.				
n	$an^3 + bn^2 + cn + d$			

EXAMPLE TABLE

n	f_n	first difference	second difference	third difference
1	6			
		18		
2	24		18	
		36		6
3	60		24	
		60		6
4	120		30	
		90		
5	210			
.				
.				
.				
n	$?n^3 + ?n^2 + ?n + ?$			

When a sequence gives a common difference after three subtractions, its nth term is cubic (is a third-degree expression). In this example, the sequence 6, 24, 60, 120, 210, ... has a common difference of 6 after three subtractions.

To find the nth term, you must find values for a, b, c, and d. Any of the expressions in the general table along with their corresponding values from the example table could be used. It is easier to solve each equation if the first expression in each column from the general table is used.

$$6a = 6$$
$$12a + 2b = 18$$
$$7a + 3b + c = 18$$
$$a + b + c + d = 6$$

Solving these equations gives $a = 1$, $b = 3$, $c = 2$, and $d = 0$. In this case, then, the nth term is $n^3 + 3n^2 + 2n$.